Occupational Health and Hygiene

Safety at Work Series*

Volume 1 – Safety Law

Volume 2 – Risk Management

Volume 3 – Occupational Health and Hygiene

Volume 4 – Workplace Safety

*These four volumes are available as a single volume, *Safety at Work*, 5th edition.

Occupational Health and Hygiene

Volume 3 of the Safety at Work Series

Edited by
John Ridley and John Channing

BUTTERWORTH
HEINEMANN

OXFORD AUCKLAND BOSTON JOHANNESBURG MELBOURNE NEW DELHI

Foreword

Frank J. Davies CBE, O St J, *Chairman, Health and Safety Commission*

My forty years experience of working in industry have taught me the importance of health and safety. Even so, since becoming Chairman of the Health and safety Commission (HSC) in October 1993, I have learned more about the extent to which health and safety issues impact upon so much of our economic activity. The humanitarian arguments for health and safety should be enough, but if they are not the economic ones are unanswerable now that health and safety costs British industry between £4 billion and £9 billion a year. Industry cannot afford to overlook these factors and needs to find a way of managing health and safety for its workers and for its businesses.

In his foreword to the third edition of this book my predecessor, Sir John Cullen, commented on the increasing impact of Europe in the field of health and safety, most notably through European Community Directives. We have since found this to be very much so. I believe that the key challenge health and safety now faces is to engage and influence the huge variety of businesses, particularly small businesses, and to help them manage health and safety more effectively. I would add that the public sectors, our largest employers these days, also should look at their management of health and safety to ensure they are doing enough.

Many businesses are willing to meet their legal obligations if given a gentle prompt and the right advice and HSC is very conscious of the importance of having good regulations which are practicable and achievable.

It is, of course, vital and inescapable that an issue as critical as health and safety should be grounded in sound and effective legislation.

This book covers many of these and other important health and safety developments, including environmental and industrial relations law which touch on this area to varying degrees. I welcome the contribution it makes towards the goal of reaching and maintaining effective health and safety policies and practices throughout the workplace.

Preface

Health and safety is not a subject in its own right but is an integration of knowledge and information from a wide spectrum of disciplines. *Safety at Work* reflects this in the range of chapters written by experts and in bringing the benefits of their specialised experiences and knowledge together in a single volume.

While there is a continuing demand for a single volume, many managers and safety practitioners enter the field of safety with some qualifications already gained in an earlier part of their career. Their need is to add to their store of knowledge specific information in a particular sector. Equally, new students of the subject may embark on a course of modular study spread over several years, studying one module at a time. Thus there appears to be a need for each part of *Safety at Work* to be available as a stand-alone volume.

We have met this need by making each part of *Safety at Work* into a separate volume whilst, at the same time, maintaining the cohesion of the complete work. This has required a revision of the presentation of the text and we have introduced a pagination system that is equally suitable for four separate volumes and for a single comprehensive tome. The numbering of pages, figures and tables has been designed so as to be identified with the particular volume but will, when the separate volumes are placed together as a single entity, provide a coherent pagination system.

Each volume, in addition to its contents list and list of contributors, has appendices that contain reference information to all four volumes. Thus the reader will not only have access to the detailed content of the particular volume but also information that will refer him to, and give him an overview of, the wider fields of health and safety that are covered in the other three volumes.

In this way we hope we have kept in perspective the fact that while each volume is a separate part, it is only one part, albeit a vital part, of a much wider spectrum of disciplines that go to make occupational health and safety.

John Ridley
John Channing
October 1998

Contributors

Dr A.R.L. Clark, MSc, MB, BS, MFOM, DIH, DHMSA
Dr T. Coates, MB, BS, FFOM, DIH, DMHSA
Nick Cook, MRSC, CChem
Occupational Hygienist, Kodak Ltd
Jonathan David, BSc
The Chartered Institution of Building Services Engineers
Frank S. Gill, BSc, MSc, CEng, MIMinE, FIOSH, FFOM(Hon),
Dip. Occ. Hyg.
Consultant ventilation engineer and occupational hygienist
Dr Chris Hartley, PhD, MSc, MIBiol
Senior Lecturer, Health and Safety Unit, Aston University
Edwin Hooper
Ron W. Smith, BSc(Eng), MSc(Noise and Vibration)
Ecomax Acoustics Ltd
Dr A.D. Wrixon, DPhil, BSc(Hons)
Principal Scientific Officer, National Radiological Protection Board

In his work, the safety adviser may be called upon to recommend measures to overcome health problems that have been identified by the doctor or nurse. Part of his duties may include the identification of processes and substances that are known to give rise to health risks and advising on the procedures to be followed for their safe use.

The advice he can give will be more pertinent if the safety adviser has an understanding of the nature of the substance and the manner in which it affects the functioning of the human body.

This book explains the functions of the major organs of the body, considers the characteristics and hazards of a range of commonly used substances and processes and discusses the techniques that can be employed to reduce the effect of these risks on the health and well-being of the workforce.

Chapter 1

The structure and functions of the human body

Dr T. Coates

1.1 Introduction

Occupational medicine is that branch of medicine concerned with health problems caused by or manifest at work. Some health problems, although not caused by the job, may be aggravated by it.

A knowledge of the structure and functioning of the organs and tissues of the body is of value in the understanding of occupational illness and injury.

Some substances are particularly liable to damage certain organs; e.g. hydrocarbon solvents may affect the liver, cadmium may damage the lungs or kidneys and mercury may affect the brain.

A brief description of anatomy and physiology is given below and more details may be obtained from textbooks on the subject.

1.2 History

Although many hazards of work were well recognised in ancient times, very little was done to prevent occupational disease. Mining was a dangerous unpleasant occupation performed by slaves. The latter were expendable and the frightful conditions in which they worked may have been a deterrent to slaves on the surface!

By the second century AD some miners were using bladders to protect themselves from dust inhalation. (Apart from armour and shields this is probably the first example of protective clothing worn at work.)

Little is known about occupational diseases in the dark ages but by the sixteenth century there was extensive mining for metals in central Europe and several accounts of associated diseases. The year 1556 saw the publication of a work of 12 books on metal mining by a mining engineer and doctor called Agricola. The latter part of book VI was devoted to the diseases of miners. Agricola advised the use of loose veils worn over the face to protect the miner against dust and ventilating machines to purify the air.

Eleven years later another doctor with an interest in mining published a work on diseases of mining and smelting. Paracelsus was physician to an Austrian town and local metallurgist. He used several metals including lead, mercury, iron and copper to treat diseases. He described the signs and symptoms of mercury poisoning and recommended the use of mercury in treating syphilis. When challenged that some of his drugs were poisonous he replied 'All things are poisons, for there is nothing without poisonous qualities. It is only the dose which make things a poison'.

In 1700 a book on trade diseases was published by an Italian physician by the name of Bernardino Ramazzini. He based the book on personal observations in the workshops of Modena where he was professor of medicine and on the writings of earlier doctors. Ramazzini was the first person to advise that physicians should ask specifically about the patient's occupation when diagnosing illness.

The development of the factory system saw the rapid movement of people from the country to the towns with consequent disruption of family life. Large numbers of workers and their families housed near the factories resulted in overcrowding, poor housing and poor sanitation. At work, people suffered appalling injury and disease and worked very long hours until eventually the pressures of humanitarians such as the Earl of Shaftesbury promoted legislation which improved working conditions and reduced the hours of employment of workers in factories, mines and elsewhere.

During this time Charles Turner Thackrah, a doctor from Leeds, wrote a book about occupational diseases in his native city which was the first such work to be published in this country. But this was 1832 and his work raised little interest, but did influence the House of Commons on future factory legislation.

The Factories Act of 1833 saw the appointment of Factory Inspectors and the need for doctors to certify that a child appeared to be at least nine years of age before being employed in textile mills. When birth certification was introduced in 1837 the assessment of children's ages became unnecessary. In 1844, the Factory Inspectors appointed Certifying Surgeons and by 1855 they were required to investigate industrial accidents and to certify that young persons were not incapacitated by disease or bodily infirmity.

By the mid nineteenth century the Registrar General had amassed a great deal of statistical information about occupational disease. Dr E.H. Greenhow of St Thomas' Hospital showed from these figures that much of the chest disease in certain areas of the country was due to the inhalation of dust and fumes at work.

In 1895, poisoning by lead, phosphorus and arsenic and cases of anthrax became notifiable to the Factory Inspectorate. Certifying surgeons examined workers in match factories, lead paint works, tri-nitrobenzene explosive factories and india-rubber factories using the vulcanising process which involved carbon bisulphide. The widespread occurrence of 'phossy-jaw' in phosphorus workers and lead poisoning gained much publicity and provoked the appointment in 1898 of Dr Thomas Legge as the first Medical Inspector of Factories.

Legge devoted the next 30 years to investigating and preventing occupational disease. His book *Industrial Maladies* was published post-humously in 1934.

By 1948, the Certifying Surgeons had become 'Appointed Factory Doctors' and numbered over 1800. They examined young people under the age of 18, investigated patients suffering from notifiable diseases and carried out periodic medical examinations on people employed in specific dangerous trades. The Appointed Factory Doctor system was replaced by the Employment Medical Advisory Service in 1972. This service, the nucleus of which was formed by the medical branch of the factory inspectorate, gives advice to employers, employees, trade unions and others on medical matters related to work.

Occupational Health Services in private industry were slow to develop and although there are rare instances of medical services at work even before the industrial revolution the first Workman's Compensation Act of 1897 was the first real stimulus which provoked employers to seek medical advice in their factories. At that time, the main reason for such appointments was to help protect the firm against claims for compensation. Exposure to hazards in munitions factories in World War I initiated many new medical and nursing appointments and increased the number of trained first aiders. Although the depression of the 1920s reversed the trend, interest returned in the 1930s and 1935 saw the founding of the Association of Industrial Medical Officers with some 20 members. This organisation grew into the Society of Occupational Medicine with a current membership of almost 2000 doctors.

A new surge of growth in Occupational Health Services occurred in World War II. The large factories were required to have their own doctors. After the war medical services grew but slowly. Many larger industries developed comprehensive medical services with X-ray, laboratory and other facilities. Some smaller factories shared medical services with their neighbours in schemes set up by the Nuffield Foundation.

In 1978, the Royal College of Physicians of London established a Faculty of Occupational Medicine as an academic centre for the subject. The Faculty has established criteria for the training and examination of specialists in the field and has a membership of over 1700.

Meantime, occupational health nursing has developed as an important aspect of health at work. Many factories with no occupational health physician employ one or more occupational health nurses. The first such nurse was employed by Colemans of Norwich in 1877. The Royal College of Nursing has recently formed a Society of Occupational Health Nursing for members employed in industry and commerce and provides training courses for those engaged in this branch of nursing. The House of Lords produced a report on Occupational Health and Hygiene Services in 1984. The report recommended development of group services which would benefit the smaller companies and suggested a Government-financed fund administered by HSE to initiate such services.

In the past decade the National Health Service has developed occupational health services for its own staff. These services are organised by individual NHS Trusts rather than on a national basis but many of them are extended to local authorities and local industry.

1.3 The functions of an occupational health department

These fall into clinical and advisory categories.

Health assessments

1 Pre-employment and other medical examinations, e.g. employees returning from sickness or those changing jobs.
2 Examination of people exposed to specific occupational hazards.
3 Treatment of conditions on behalf of the hospital or general practitioner. This may include physiotherapy or rehabilitation for which purposes a physiotherapist may be employed.
4 Emergency treatment of illness or injury occurring at work.
5 Immunological services, e.g. vaccination of overseas travellers, tetanus prevention, influenza prevention. Hospital workers require protection against hepatitis and tuberculosis.

Advisory services

1 The study and prevention of occupational disease.
2 Advice on problems of medical legislation and codes of practice.
3 Advice on medical aspects of new processes and plant.
4 The study of sickness absence.
5 Advice on the reduction or prevention of common non-occupational diseases such as alcoholism and the effects of smoking.
6 Training first aiders.
7 Advice to employees prior to retirement.
8 The preparation of contingency plans for major disasters at the place of work.

Nurses have an important part to play in these activities and much of the clinical treatment of patients is in their hands. Nurses may be State Registered (SRN) or Registered General Nurse (RGN) with 3 years' training or State Enrolled (SEN) with 2 years' practical training. Full-time and part-time training courses in occupational health nursing are run by the Royal College of Nursing at various centres. The RGN may obtain a diploma in occupational health nursing after an examination. The SEN may take part in one of the courses which will help her carry out her duties in this field of nursing. As most nurses in industry and commerce lack full-time medical advice the need for formal training in the subject is very clear.

1.4 Overseas developments

Not all EC countries have introduced legislation on occupational health. In France, for example, there is no law requiring treatment services but pre-employment medical examinations are mandatory.

Holland and Belgium require medical services in companies of over a specified size. In Germany, doctors trained in occupational health must be employed in factories as must safety advisers, and in a number of other European countries the major concerns have occupational health services.

Some countries use the factory as the site for a medical centre which provides clinical services for workers and their families as well as making available similar medical facilities to those provided by many factory medical departments in this country.

In the USA the National Institute for Occupational Safety and Health (NIOSH) determines standards of occupational health and safety at work and organises training and research facilities. The Occupational Safety and Health Act 1970 applies to workers in industry, agriculture and construction sites and requires that employers must provide a place of work free from hazards likely to cause death or serious harm to employees[1].

1.5 Risks to health at work

The main hazards are of three kinds, physical, chemical and biological, although occupational psychological factors may also cause illness.

1.5.1 Physical hazards

Noise, vibration, light, heat, cold, ultraviolet and infrared rays, ionising radiations.

1.5.2 Chemical hazards

These are liable to occur as a result of exposure to any of a wide range of chemicals.

Ill-effects may arise at once or a considerable period of time may elapse before signs and symptoms of disease are noticed. By this time the effects are often permanent.

1.5.3 Biological hazards

These may occur in workers using bacteria, viruses or plants or in animal handlers and workers dealing with meat and other foods. Diseases produced range from infective hepatitis in hospital workers (virus infection) to ringworm in farm labourers (fungus infection).

1.5.4 Stress

This may be caused by work or may present problems in the time spent at work. Work related stresses may be due to difficulties in coping with the amount of work (quantitative stress) or the nature of the job (qualitative stress).

1.6 Occupational hygiene

In 1959, the American Industrial Hygiene Association defined occupational hygiene as 'that science and art devoted to the recognition, evaluation and control of the environmental factors or stresses arising in or from the workplace which may cause sickness, impaired health and well-being or significant discomfort and inefficiency among workers or among citizens of the community'[2].

The British Occupational Hygiene Society was founded in 1953 'to provide a forum in which specialist experience from many different but related fields could be exchanged and made available to the growing number of occupational hygienists at both national and international level, and to encourage discussion with other managerial and technical professions'. The Society holds frequent conferences and meetings.

The British Examining Board in Occupational Hygiene was set up by the British Occupational Hygiene Society in 1968 to examine candidates to well-defined professional standards.

The first stage in the practice of good occupational hygiene is to recognise the potential or manifest hazard. This may result from an inspection of the process in question or may be suggested by symptoms and signs of disease in the operatives. Ideally the potential risk should be considered at the planning stage before plant is installed.

The next stage is to quantify the extent of the hazard. Measurements of physical and chemical factors and their duration must be related to levels of acceptability and the likelihood of injury or disease arising if the hazard is allowed to continue. These measurements often involve the use of sophisticated measuring devices which must be calibrated and used very carefully in order to produce meaningful results.

For smaller firms or small isolated units in larger organisations, the person who carries out a limited range of tests may benefit from attendance at a short course leading to a Preliminary Certificate in Occupational Hygiene at a College of Further Education. Full-time specialists in Occupational Hygiene will require a professional qualification in the subject. The need to meet the requirements contained in the Control of Substances Hazardous to Health Regulations 1994 has increased the role of occupational hygienists, both full-time and part-time.

Having assessed the dangers of the process, the final stage is to decide how best to control the hazard. This may require some radical modification of plant design, special monitoring devices which will warn of increasing danger or the need for protective devices to be used by plant operators.

In deciding on appropriate ways of dealing with such problems, the occupational hygienist will often require the co-operation and understanding of the occupational physician, nurse, safety adviser, personnel officer and line management in order to achieve his aims.

The degree of involvement and co-operation of advisers in the medical, nursing, engineering and safety fields will vary from one problem to another. Failure to achieve adequate health and safety measures may be due to lack of understanding or co-operation between advisers in the various disciplines or failure to influence line management.

The appointment of safety representatives under the Health and Safety at Work Act 1974 has focused attention on the part played by all who are involved in occupational health, safety and hygiene. Advice on prevention and safety measures is less likely to be ignored but more likely to be challenged than in the past. It is vital that the adviser's opinions can withstand challenge and are seen to be fair and unbiased.

1.7 First aid at work

The obligations placed on an employer to provide first aid facilities are contained in the Health and Safety (First Aid) Regulations 1981[3]. These Regulations recognise that the extent to which first aid provision is required in the workplace depends on a range of factors including hazards and risks, the size of the organisation, the distribution of the workforce, the distance from various emergency services and the extent of the occupational health facilities provided on site. When doctors and/ or nurses are employed this can be a factor in determining the number of first aiders needed.

An assessment of the risk to employees' health and safety is required under MHSW and this should include an assessment of first aid requirements. If the risk is small, an easily identified fully equipped first aid materials container and an *appointed person*, trained to deal with emergencies, may be all that is required. Short courses lasting about 4 hours are available for training such people to cope with emergencies and the trainers do not require HSE approval.

In areas of greater risk, the employer needs to provide better services. In such areas a first aider should be available to give first aid immediately after an accident has occurred. Where the process includes the possibilities of gassing or poisoning, the first aider may need special training to deal with these specific risks. Adequate first aid rooms and equipment may be required, especially in areas of high risk such as the chemical industry and on construction sites.

Details of all first aid treatments should be recorded. The record may be made in the statutory accident book (BI 150) or in a record system developed by the employer. The local emergency services should be notified of all sites where hazardous substances are used.

1.7.1 Number and qualifications of first aiders

These should be available in sufficient number to be able to give first aid rapidly when the occasion demands. Where more than 50 people are employed a first aider should be provided unless the assessment can justify other facilities. For example, a small organisation with only minor hazards may require only an appointed person. On the other hand, serious hazards may warrant a first aider in each hazardous area. A detailed guide to the assessment of first aid needs is given in an Approved Code of Practice[4] which includes a table indicating the recommended number of first aiders.

First aiders must hold a currently valid certificate of competence in first aid at work. The HSE approves first aid trainers and provides information on the availability of local courses. Certificates obtained after basic training last for 3 years but refresher courses should be attended on a regular basis particularly leading up to the renewal of certificates. Where special training has been given for specific hazards, the standard certificate may be endorsed to confirm that such training has been received.

1.7.2 First aid containers

The contents of a first aid container should match the assessed needs of the work area, the lower the risk, the more basic the contents. It is the responsibility of managers to ensure that they provide sufficient number of containers, reflecting the extent and location of the hazards, and that the contents of each container are appropriate to the risks faced in the work area covered. Tablets and medication should not form part of their contents.

First aid containers should be sufficiently robust to protect the contents and should be clearly identified by means of a white cross on a green background. The containers should contain only first aid items, including a guidance leaflet[5], and should be kept properly stocked. This may require the holding of back-up first aid supplies. The approved Code of Practice[4] suggests the minimum content for a container in a low risk work area. First aid containers should be located near hand washing facilities. If peripatetic workers have to visit areas without first aid cover, they should be provided with a small travelling first aid kit.

1.8 Basic human anatomy and physiology

Anatomy is the study of the structure of the body. Physiology is concerned with its function. Although the various organs which make up the body can be studied individually it is important to remember that these organs do not function independently but are interrelated so that if one part of the body is not functioning properly it may upset the health of the body as a whole.

An organ like the stomach or the brain contains structures within it such as arteries, nerves and other specialised components. These components, which contain cells of a similar kind, are referred to as tissues. So we have nervous tissue, arterial tissue, muscular tissue and so on making up specialised organs which have a specific function. (The stomach, for example, is concerned with the first stage in the digestive process.)

The cells which make up the tissues and organs are so small that they are invisible to the naked eye. Under the microscope a cell consists of a mass of jelly-like material called protoplasm held together by a surrounding membrane. The shape and function of the cell vary according to the tissue of which it is composed and depend upon the job which the cell is required to do. For example, the nerve cell has long fibres capable of conducting electrical impulses while some cells in the stomach wall produce hydrochloric acid. Cells in the thyroid gland produce a chemical which influences other body cells. With these varying roles it is not surprising that cells differ from one another in appearance.

Although the human body is composed of many million cells, the work of each one is controlled so as to serve the body as a whole. If this coordination is lost, some cells can grow rapidly relative to others and the result may be disastrous. This sort of cell behaviour occurs in cancer when a group of cells may grow rapidly invading adjacent tissues.

Each cell is a sort of miniature chemical factory. It takes in food and converts it into energy to perform work. The sort of work carried out depends on the type of cell, e.g. locomotion (muscle cells), oxygen transport (trachea, lungs, blood vessels, red blood cells). Energy is also needed to repair the wear and tear of body cells. We refer to the chemical processes which convert food into energy as metabolism.

1.8.1 Foodstuffs

The energy needed to perform work and to maintain body temperature is provided by oxygen and various foods. Any diet which maintains life must contain six basic ingredients in a digestible form. These are as follows:

1 Proteins.
2 Carbohydrates.
3 Fats.
4 Salts.
5 Water.
6 Vitamins.

Proteins are composed of large complicated molecules which contain atoms of carbon, hydrogen, oxygen, nitrogen and often sulphur. They are made up of simpler substances called amino acids which form the basic structures of body cells. Foods such as meat, milk, beans and peas contain protein. This is broken down in the digestive process into its constituent aminoacids which are then realigned in a new pattern to make human proteins.

As the name suggests, *carbohydrates* are composed of carbon, hydrogen and oxygen. Sugars and starches are common examples of carbohydrate. *Fats* are used as reserve foodstuffs and insulate the body thus protecting it against cold. Carbohydrates and fats are a ready source of heat energy.

Various *salts* including those of sodium, iron and phosphorus are obtained from food such as milk and green vegetables. They are needed for the formation of bone, blood and other tissues.

Water is a vital constituent of all cells and a regular intake is essential to maintain life. Small quantities of a range of chemicals known as *vitamins* are also needed for healthy existence. The absence of a vitamin leads to a deficiency disease such as scurvy which occurs when vitamin C is absent from the diet. It is available in fresh vegetables, oranges and lemons. Vitamin D is formed by the action of ultraviolet light on a chemical in the skin (7-dehydrocholesterol), and is present in milk and cod liver oil. Absence of this vitamin may lead to rickets.

As well as containing the substances listed above an adequate diet must provide enough calories to satisfy metabolic requirements. This will depend on the physical demands of the person's work and hobbies as well as on his stature.

1.8.2 Digestion

When food is taken into the body much of it is in a form which cannot be used directly by the tissues as its chemical structure is too complicated. The larger molecules of food therefore need to be broken down into simpler molecules. This process takes place by chemical action and occurs in the digestive tract (*Figure 1.1*) which is a long tube of varying dimensions which starts at the mouth from where the food passes to the gullet, the stomach, the small intestine and finally the large intestine.

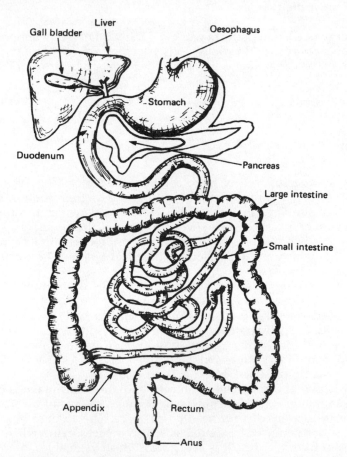

Figure 1.1 Diagram of digestive system

Alcohol is absorbed in the stomach and water is absorbed in the large intestine but the majority of energy-containing foods are absorbed in the small intestine. The products of digestion pass through the walls of the digestive tract into blood vessels and thence to the liver. This is a very large organ situated in the upper right side of the belly cavity below the diaphragm. It is made up of many units of cells arranged around blood vessels in a circular fashion. Sugar from the digestive tract is taken to the liver where it is changed into a chemical called glycogen which is stored in the liver cells. Proteins are broken down in the liver and form urea as a waste product which is then excreted via the kidneys and is the main chemical constituent of urine.

Old red blood corpuscles are removed from circulation by the liver which retains iron from them for later use. The liver is also responsible for the manufacture of bile which assists digestion and which is stored in the gall bladder adjacent to the liver.

Many poisons are dealt with by the liver which attempts to render the poison less toxic (detoxification) before it is excreted. Sometimes the poison damages or destroys liver cells but fortunately the liver has such a large reserve of cells that a great deal of damage is necessary to affect its function adversely. Liver damage may result from certain types of industrial poisons as well as from excessive consumption of alcohol.

1.8.3 Excretion

Just as a motor car needs to get rid of exhaust gases so the human body has to dispose of waste materials left over from metabolic chemical reactions. Special organs are involved in the process of excretion. Water and urea are disposed of by the kidneys, although some water is lost via the skin. Solid waste leaves the body through the bowel after water has been extracted in the large intestine.

The voiding mechanisms are of paramount importance when the body is affected by poisonous materials and may be by vomiting, diarrhoea or by being excreted in the urine. Sometimes the passage of a poison through the body may leave a trail of destruction in its wake and result in permanent liver or kidney damage.

1.8.4 The respiratory system

The foodstuffs absorbed from the digestive tract are converted into energy. In order to produce this energy the body cells require oxygen just as a motor car engine needs oxygen in order to function. This converting process generates carbon dioxide which, in large quantities, is poisonous and must be got rid of.

During the process of respiration oxygen is transferred from the air to the body cells and carbon dioxide is disposed of in exhaled air. Because body cells function at different rates their oxygen requirements vary from one tissue to another. If brain cells are starved of oxygen for more than four minutes there is little prospect of recovery of intellectual function. Other body cells can do without oxygen for longer periods of time.

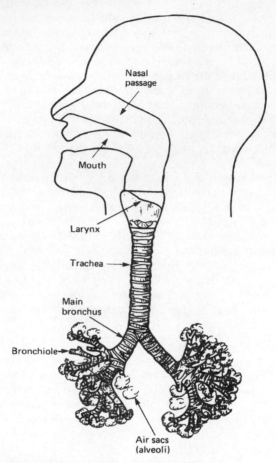

Figure 1.2 Respiratory system

A great deal of oxygen is needed to provide the requirements of all the body cells. The apparatus which fulfils this requirement is the respiratory tract (*Figure 1.2*) which is made up of the nose and mouth, the throat, the larynx (voice box) and trachea (windpipe), the bronchi and the lungs. The respiratory tract is lined by a wet shiny membrane which contains mucous secreting cells that keep the walls moist. Other cells are fringed with little hairs or cilia which by moving in one direction can evict towards the mouth particles of dust which have entered the airways. Sometimes the quantity of material which has entered the airway is greater than the cilia can cope with. The lungs then eject collections of particles by the mechanism of coughing. Primary filtration of air entering the respiratory tract occurs in the nose but many of the smaller particles enter the air passages where some hit the walls of the bronchi and are rejected by the cilia but others go on to reach the lungs.

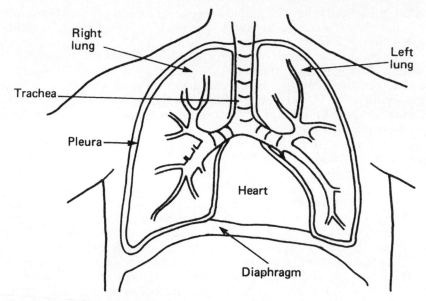

Figure 1.3 Diagram showing lungs and heart within chest

The respiratory tract is rather like an inverted tree. The windpipe is the trunk and the two main bronchi (one for each lung) the main branches. Each main bronchus then subdivides into smaller branches (bronchioles). All these tubes are held open by rings of cartilage which prevent their collapse when subject to suction when breathing in. The smallest tubes end in air sacs or alveoli which have very thin walls. These walls allow oxygen to pass into the small blood vessels with which they are surrounded and carbon dioxide to pass in the opposite direction from the blood into the air sac.

The blood vessels lining the air sacs carry the oxygen-containing blood to the left side of the heart from whence it is pumped via the arteries to all parts of the body. When the tissues have used the oxygen to carry out the metabolic processes described earlier carbon dioxide is produced which like oxygen is dissolved in the blood. The carbon dioxide is carried back to the lungs via the veins and the right side of the heart.

The lungs are surrounded by a tough layer of smooth membrane called the pleura which if it becomes inflamed gives rise to a condition known as pleurisy. Blue asbestos fibres can irritate the pleura to produce a type of cancer called a mesothelioma.

The lungs are located in the chest cavity in a space limited by the ribs, breastbone, backbone and diaphragm (*Figure 1.3*). The latter is a dome-shaped sheet of muscle which separates the chest and belly cavities. When the diaphragm moves downward the dome shape is flattened and at the same time the ribs move upwards, thus increasing the volume of the chest cavity creating a suction which draws air into the lungs and then into the air sacs (about 20% of air is oxygen). This movement is called inhalation. When the diaphragm expands the chest cavity contracts and the elastic recoil of the lungs forces air in the opposite direction (exhalation).

If respiratory movement ceases (e.g. due to electric shock damaging the breathing mechanism) artificial respiration is needed at once. This may be carried out by breathing into the casualty's mouth (mouth to mouth resuscitation) or by exerting pressure on the casualty's chest, thereby forcing air out of the lungs and allowing the recoil of the chest wall to draw air into the lungs.

Divers and caisson workers may suffer from pain in the joints if they return to the surface too quickly after working in deep water or under elevated air pressure. In these conditions, air contained in the body tissues, which was dissolved under high pressure, is released to form air bubbles (mostly nitrogen gas) in the joints and elsewhere to produce unpleasant symptoms. 'The bends' is the name of the illness resulting. Mild cases affect the elbows, shoulders, ankles and knees. As the illness develops, the pain increases in intensity and the affected joint becomes swollen. Serious cases of the bends may affect the brain and/or the spinal cord. In cases of brain damage the patient may suffer visual problems, headaches, loss of balance and speech disturbances. Spinal cord damage may cause paralysis of the limbs, loss of sensation, pins and needles and pain in the shoulders and/or hips. The problem is obviated by reducing the rate of change of pressure to which the workers are subjected to a level at which the bubbles of gas do not form.

1.8.5 The circulatory system

This consists of the heart, the arteries, the veins and the smaller blood vessels which permeate all tissues of the body (*Figure 1.4*). The heart is a muscular pump divided into left and right sides. The left side is larger and stronger than the right since it has the bigger job to do in pumping to all parts of the body via the arteries oxygen-containing blood which it has received from the lungs. The blood then takes carbon dioxide from the tissues and carries it back to the lungs through the veins via the right side of the heart.

Each side of the heart (*Figure 1.5*) has two chambers, an auricle or intake chamber and a ventricle or delivery chamber which are separated by valves which ensure that the blood travels in one direction only from auricle to ventricle. A further set of valves ensures that the blood being ejected cannot flow back into the ventricles each time the heart contracts.

The muscles of the heart are less dependent on the brain than the muscles which cause body movement. Cutting the nerves to a muscle in the leg, for example, will stop that muscle working. Cutting the nerves to the heart will not stop the heart since it is not entirely controlled by nervous impulses coming from outside the organ as is the case with voluntary muscle. The heart has a dual nerve supply. One nerve supply increases the heart rate, the other reduces it. The rate of the heart beat,

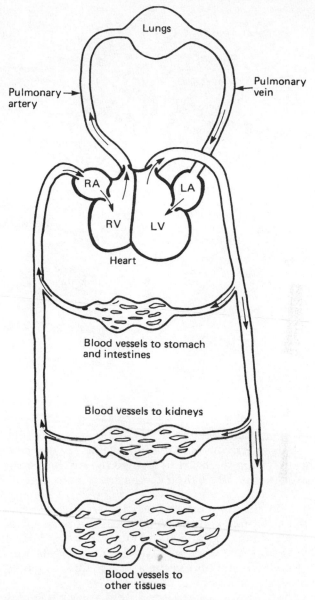

Figure 1.4 Diagram showing circulation of blood (RA – right auricle; LA – left auricle. RV – right ventricle; LV – left ventricle.)

which in the healthy adult at rest is about 70 beats per minute, may be slowed by stimulating the vagus nerve and increased by stimulating the sympathetic nerves. The latter are activated when people are afraid so that more blood is sent to the muscles and the person is keyed up for action.

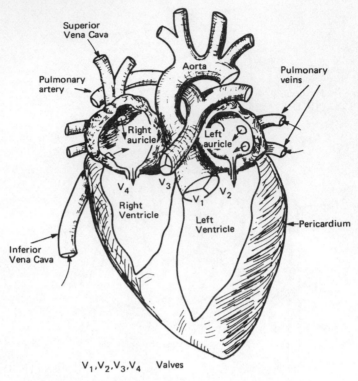

Figure 1.5 Diagram of human heart

Electric shock may affect the heart rate either by stimulating the nerves to the heart or by interfering with the conduction of electricity to the heart muscle. This may stop the heart or change the rhythm thus affecting its efficiency as a pump. Fibrillation of the heart muscle may occur as a result.

In addition to its oxygen-carrying capacity, the blood also carries nutrients to the tissues and removes waste products. It enables heat generated to be dissipated and its white cells defend the body when attacked by viruses and bacteria. The blood accounts for about one-thirteenth of body weight and in the average adult amounts to about 5 litres (12 pints).

It is composed of a straw-coloured fluid (plasma) and cells of two different colours – red and white. The red blood cells which account for the colour of blood are made up of minute circular discs which contain red pigment (haemoglobin) with which the oxygen and carbon dioxide transported in the blood combine temporarily. Haemoglobin also combines very readily with carbon dioxide.

The white blood cells are somewhat larger than the red blood cells but fewer in number. There is only one white cell for every 500 red cells. Several kinds of white cell exist which are mobilised when the body is infected by germs and viruses and attempt to destroy them.

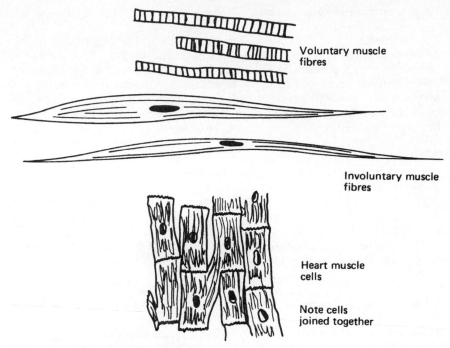

Voluntary muscle
fibres

Involuntary muscle
fibres

Heart muscle
cells

Note cells
joined together

Figure 1.6 Structure of three types of muscle

1.8.6 Muscles

One of the characteristics of all animals is movement. In man, movement is brought about by the contraction of muscles. There are three varieties of human muscle (*Figure 1.6*). Voluntary muscles which can be moved at will are made up of many fibres which are covered by transverse stripes when seen under the microscope. The muscle is often attached to bones (hence the term skeletal muscle) by connective tissue fibres which collectively form a tendon. When the muscle contracts, the bones to which it is attached are drawn together producing movement of one part of the body relative to another part.

Involuntary muscle does not have the characteristic striped appearance of voluntary muscle. It exists in the digestive tract, the walls of the blood vessels and in the respiratory and genito-urinary apparatus. Involuntary muscle is controlled automatically by the autonomic nervous system.

The third type of muscle is that found in the heart. This shows some striated fibres under the microscope but is not under voluntary control and is not entirely dependent on its nerve supply. A 'pacemaker' within the heart muscles produces, at a rate appropriate to the body's need for oxygen, electrical impulses which cause the heart muscles to contract thus producing heart beats.

1.8.7 Central nervous system

This is made up of the brain and spinal cord. The brain is a highly developed mass of nerve cells at the upper end of the spinal cord. The largest part of the brain is taken up by the two cerebral hemispheres. These receive sensory messages from various parts of the body and originate the nerve impulses which produce voluntary movements. The layers of grey tissue (known as the cerebral cortex) overlying the cerebral hemispheres are covered in folds. This tissue is concerned with the intellectual function of the individual. The various parts of the cortex of the brain are associated with specific activities. For example, there are centres concerned with speech, hearing, vision, skin sensation and muscle movement.

The cerebellum is the part of the brain concerned with balance and complicated movements. Part of the base of the brain is involved with emotional behaviour (the hypothalamus).

The portion of the brain nearest the spinal cord contains centres which control the rate of respiration and heart beat.

1.8.8 The special senses

These specialised organs measure environmental factors such as light and noise and facilitate communication with other human beings.

The nerve cells within these organs pass their messages to the brain which then interprets them and determines appropriate action.

1.8.9 The eye

The eyeball (*Figure 1.7*) is a globe 25 mm (1 inch) in diameter which is made up of a transparent medium (the vitreous) through which light is

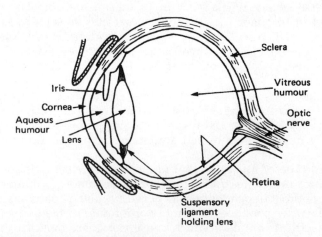

Figure 1.7 Diagram of section through eyeball

focused by a lens onto a light sensitive layer (the retina). The front of the globe (the cornea) is transparent. After light rays have entered the eye they pass through a fluid (the aqueous humour) and the lens which changes in shape in order to focus on to the retina light rays from objects at varying distances.

If the path of light rays is interrupted, e.g. by a foreign body or by an opacity of the lens (cataract), the vision may be distorted and/or diminished.

Light produces changes in the cells of the retina which transmit electrical impulses to the visual cortex at the back of the brain. The movements of the eyeball are controlled by six muscles in each eye which are carefully synchronised with those of the other eye. Imbalance of these muscles may give rise to double vision or a squint. Temporary muscle imbalance may result from exposure to toxic materials or alcohol.

Burns of the eye may result from chemical splashes and exposure to ultraviolet radiation as can occur with electric welding. It is vital that chemical burns be treated at once by irrigation with copious quantities of running water.

1.8.10 The ear

This organ is concerned with hearing and with orientating our position in space. The portion concerned with hearing consists of three parts (*Figure 1.8*). Sound travels through the ear canal to the eardrum which is a membrane stretched across the canal and separating it from the middle ear. Vibrations of the eardrum are produced by the sound waves passing along the ear canal.

These vibrations are transmitted through the middle ear by three tiny bones known as the 'ossicles', being the hammer (malleus), anvil (incus) and stirrup (stapes). The hammer bone is fixed to the eardrum and the stirrup to another membrane (the oval window) which separates the middle and inner parts of the ear. The section of the inner ear which receives sound waves is shaped like a snail's shell (the cochlea) and contains strands of tissue under varying tensions. These strands or hairs vibrate in response to sound waves of particular frequencies which have entered the inner ear from the bones of the middle ear and produce nerve impulses in the auditory nerve which are then transmitted to the cortex of the brain. It is at this point that the signals are received as sound of a certain pitch, intensity and quality.

Various factors may interfere with the transmission of sound impulses. Normally, the pressure on either side of the eardrum is equal but when a difference occurs, as with airline passengers who fly when suffering from a cold, temporary hearing impairment can be experienced. Infection of the middle ear may occur and this may result in thickening and scarring of the eardrum. Some unfortunate people suffer an inherited form of deafness in which the ossicles develop damage and are unable to transmit sound.

The inner ear is a very sensitive part of the hearing mechanism and may be damaged by prolonged loud noise. Usually, the frequencies around 4000 hertz (cycles/second) are first affected but the damage can

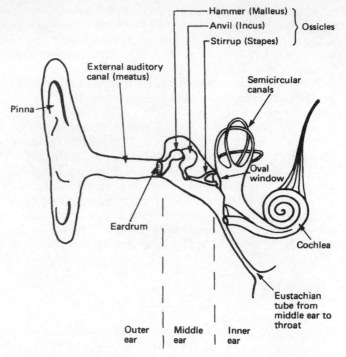

Figure 1.8 Diagram of human ear

extend to other frequencies as well as becoming more pronounced. In addition to this, deafness is associated with the ageing process and is more noticeable in males (presbyacusis).

A balance mechanism is also situated in the inner ear. This is composed of three semicircular canals which are at right angles to each other. Inside each canal are specialised nerve endings.

Moving the body into an unbalanced position stimulates the nerve endings in one of the canals in each ear and results in an urge to return the body to a normal balanced posture.

1.8.11 Smell and taste

The lining of the inside of the nose contains special cells which are capable of detecting chemicals in the air. Nerve fibres pass from these cells into the skull and connect with the brain.

The sense of smell may be an important safety factor. A cold may diminish or remove the facility. Hydrogen sulphide gas smells of bad eggs. Continuing exposure to increasing concentrations of this gas saturates the nerves concerned with smell so that the person exposed to this substance may be unable to smell it even if the concentration increases further.

The sense of taste originates when chemical stimulation of the taste buds occurs. These are collections of cells concentrated in certain areas of the tongue. The sides, tip and rear third of the organ have the most taste buds. The back of the tongue most readily detects bitterness and the tip sweetness.

1.8.12 Hormones

These are chemicals which act as messengers provoking action in some distant part of the body. They are produced by various hormone or endocrine glands. For example, the thyroid gland is a gland situated in the front of the neck which produces the chemical thyroxine. Too much thyroxine produces a rapid pulse and an overactive jumpy person. Too little thyroxine may result in a slowing of the pulse and too slow a rate of metabolism with the face becoming swollen and the skin dry and aged; the hair becomes coarse and falls out (myxoedema).

The suprarenal glands are two small glands situated above the kidneys. They produce a number of hormones including adrenaline and cortisone. Adrenaline is released in conditions causing fear or anger and makes the muscles in the artery walls contract. This increases the blood pressure, and consequently the supply of oxygen to the muscles, so that an animal is ready to meet a confrontation by either 'fight or flight'. This is not always an appropriate reaction for human beings in present day stressful situations where they cannot fight or run away.

Cortisone is released from the adrenal cortex at times of stress. It delays physical fatigue by increasing the ability of muscles to contract and has a euphoric effect on the brain which may give added confidence at a stressful time.

Stress is an engineering term describing the force applied to an object and the resulting deformity is referred to as strain. It has become customary to refer to the result of applying such force or pressure on human beings as 'stress'.

The effects of long-term stresses on the human body and mind are not clearly understood. Certain so-called 'stress diseases' such as asthma, duodenal ulcer and coronary heart disease may be aggravated at times of stress but a direct cause/effect relationship is difficult to prove.

Nevertheless, people in stressful situations, whether caused by work or by non-occupational factors, may be more liable to accidents.

Another hormone-producing organ, the pancreas, has two important functions. Its secretions flow into the digestive tract where the gland's products are involved in the digestion of carbohydrates. A different secretion which passes straight into the blood stream is a chemical called insulin. Without insulin the body is unable to use available carbohydrates as a source of energy and has to obtain energy from the breakdown of body tissues. The person whose pancreas is unable to make enough insulin for his needs is diabetic. Diabetes is a condition which can be treated by replacing the insulin deficiency and by careful dietary control.

The sex glands also produce hormone secretions which determine the growth of beard hair and the deep voice of the male and breast development in the female. Other endocrine glands include the parathyroid glands which are concerned with calcium metabolism and the pituitary gland which controls the other hormone glands. The pituitary gland situated at the base of the brain has two lobes: the front one controls growth in children while the rear lobe secretions cause contraction of the muscles of the womb and increase the output of urine. The part it plays in regulating the other endocrine glands has been referred to as 'direction of the endocrine orchestra'.

1.8.13 The skin

This is the largest organ in the body (*Figure 1.9*) and performs a variety of functions. Its most obvious purpose is a protective one. The superficial layers of cells keep out chemicals and germs as well as acting as a physical barrier. If the physical pressures on certain cells of skin are considerable, the tissues may be much thickened, for example on the soles of the feet.

Four different kinds of sensation may be appreciated through the skin, namely heat, cold, touch and pain.

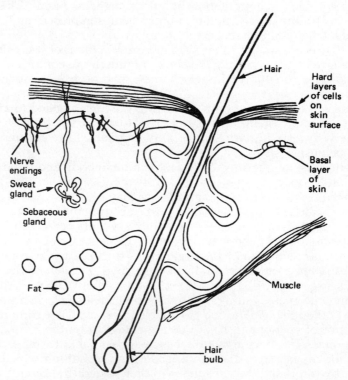

Figure 1.9 Section through skin

The skin is not uniformly sensitive to all these stimuli, some parts being more sensitive than others; for example, the tips of the fingers are more sensitive than the back of the hands. Apart from our awareness of touch and temperature on the surface of the skin, we are also aware of the position of our arms and legs even when our eyes are closed. This is because special nerve endings from the muscles return nerve impulses to the brain which is provided with information on the relative positions of various parts of the body. Water and various salts are lost through the skin as sweat. This may be important to people working in very hot environments such as in deep mines who may sweat profusely and as a result of the loss of salt suffer severe cramps in the muscles.

The human being is able to keep his body temperature fairly constant whatever the range of ambient temperature. This is because heat loss from the skin may be increased by sweating, followed by evaporation of the sweat and increase in the size of the blood vessels in the skin which encourages more heat to reach the skin surface.

In a cold environment the blood vessels in the skin contract and the muscles under the skin produce the phenomenon of shivering in an attempt to generate heat.

The skin also secretes sebum which is a waxy material produced by special glands. This may help protect the skin surface from attack by foreign substances. Exposure to various chemicals including solvents may remove such secretions and predispose the skin to attack by germs.

1.9 Cancer and other problems of cell growth

Cells reproduce in order to replace other cells which are continually being lost, e.g. on the surface of the skin by wear and tear or by damage such as that caused in a wound.

Sometimes, cells do not develop as nature intended and a variety of abnormal cells may be formed, some of which may endanger life. Some examples of maldevelopment are given below:

Aplasia   The tissues may fail to develop. This happened in the unfortunate children whose mothers had taken the drug thalidomide during their pregnancies. In some of these babies, the limbs were only partly developed (hypoplasia).

Hyperplasia   The organ or tissue has an increased number of cells. When the cells increase in size this is termed hypertrophy. The latter condition may occur in the muscles of a leg when the other has been amputated and the remaining limb has to work harder.

Metaplasia   This process involves changes in the types of cell present in a tissue often from a complicated cell to a more simple one. Long lasting irritation from chemicals may bring about this change which is often seen in the lining of the air passages of smokers.

Mutagenesis   This occurs when any abnormal changes which may have arisen in a group of cells are passed on to the next generation as the original genetic material within the cell has been damaged.

Teratogenesis   This refers to congenital abnormalities which occur in the new-born as a result of damage to the genetic material or to the developing embryo.

Neoplasia   This is the process in which there is a mass of new tissue (a tumour) which is made up of abnormal cells which cannot stop growing. Malignant tumours are more likely to occur in tissues which are frequently producing new cells.

The chemical DNA which carries the coding mechanism which determines the structure and function of new cells is liable to be damaged by some chemicals. New malformed cells may form and grow in an uncontrolled manner producing more damaged cells of the same type. Malignant tumours consist of undifferentiated cells which grow into surrounding tissues. Benign tumours are slower to develop and their cells look more like normal cells.

Cancer cells may spread and 'seed' themselves elsewhere in the body. From these new sites further secondary cancerous growths may develop, hastening the destructive process which may well end in the victim's death. Neoplasms may spread via the blood stream or along lymph vessels.

1.10 The body's defence mechanisms

The various systems of the body have a range of mechanisms for dealing with foreign materials, germs and viruses that enter the body.

Particles entering the trachea or smaller airways may be ejected by the reflex action of coughing. Irritating substances entering the nose will provoke sneezing. The lining of the nose and upper airways produce a glairy fluid (mucus) on which particles may land in which case they are dealt with by small moving hairs (cilia) which propel them towards the throat and nose. However, the smallest particles may be inhaled into the lungs where they can cause damage. Gases may be absorbed from the air sacs (alveoli) and enter the blood stream which conveys them to tissues where harm can be caused if the gases are toxic.

Foreign bodies on the surface of the eye may cause weeping (lachrymation) which can flush out some of the particles but not those that have become embedded in the eye. Tears will also dilute irritating chemicals.

The immune system provides a wide variety of defence mechanisms. Bacteria entering the blood stream are engulfed in a type of blood cell (the macrophage), a process known as phagocytosis. Macrophages in the lungs attack other foreign bodies as well as bacteria. They may also lodge in organs, such as the liver and spleen, where they eat any organism that invades their area.

A second type of white cell, the polymorphonuclear leukocyte (PMNL) is more common than the macrophage and makes up 60% of all white blood cells. These cells are attracted to sites of infection by the chemicals that are released by invading bacteria. Special cells, known as B lymphocytes, produce antibodies that are chemicals that can attach themselves to the surface of various germs thus rendering them ineffective, but any particular antibody can only bind to one bacterium. T cytoxic (killer) cells can kill cells directly. They attacked cells filled with viruses before the virus can reproduce and are involved in the defence of the body against cancer cells.

When the body is severely affected by invading organisms or irritant chemicals or burns, inflammation of the affected tissue occurs. The signs and symptoms of the inflammatory process include heat, redness, swelling, pain and loss of function of the affected part of the body. These reactions result in an increased blood supply to the affected part where the beneficial effects of the immune system's cells are enhanced. Severe inflammatory damage may be followed by scarring of the affected tissues when collagen fibres form in the damaged area. These fibres are important in would healing where they strengthen the damaged tissues. However, scarring in the lungs may reduce the efficiency of that organ and scarring within the eye may reduce visual acuity.

1.11 Factors determining the effect of substances in the body

Poisonous materials may enter the body via the mouth and be absorbed from the food tract, via the airways and lungs or through the skin.

The quantity of, and duration of exposure to, a given substance are the most important factors in determining the severity of effects on an individual. Specific chemicals can cause damage to particular organs (target organs), e.g. solvents may cause damage to the liver and kidneys. If these organs have already suffered previous damage from some other disease process, they will be less able to cope with further insult from occupational hazards.

Another factor in determining the toxicity of a substance is its solubility in body fluids or the ease with which it becomes attached to tissues within the body. Carbon monoxide owes its high toxicity to its ability to attach to molecules of haemoglobin, the pigment inside red blood cells. Some materials cause disproportionate damage by synergistic effects. Both smoking and asbestosis (the presence of asbestos particles in the lungs) are well-recognised causes of lung cancer but there is a far higher risk of this disease in smokers with asbestosis than one would expect by adding together the risk from each hazard.

Some chemicals act as 'sensitisers' provoking an unpleasant response, usually in the lungs or the skin, from a very small quantity of the offending substance. This process is caused by a response of the body's immune system to foreign material.

In the case of particle inhalation, their size and shape may determine whether they are retained in the lungs or are rejected. Long thin needle-like fibres, such as asbestos, may be more readily retained than rounder particles, and may cause more physical irritation.

Some materials have an immediate adverse effect on surrounding tissue. This so-called acute effect is seen when strong acids or alikalis are in contact with body tissues. Other substances take much longer, perhaps weeks, months or years, before adverse effects are noticed, e.g. cancers. These are termed chronic effects which, unfortunately, tend to be irreversible changes that often do not respond well to treatment.

1.12 The assessment of risk to health

The hazards of specific substances are described in textbooks on toxicology and the details of hazardous substances used in industrial processes may be obtained from the suppliers' safety data sheets. However, the employer is responsible for carrying out assessments of the risks that may arise from the use of those substances in the particular circumstances of his own plant.

Various systems of the body can be monitored to identify the presence of harmful substances so that people exposed to a known hazard may be removed from contact with the offending agent before clinical damage ensues. Examples of such monitoring include sampling of the blood and urine for chemicals, lung function tests to assess the efficiency of the lungs and the measurement of visual and hearing acuity. When the functions of specific organs are being measured it is important to note that non-occupational disease or previous ocupational exposure may account for some or all of the damage revealed by such tests.

References

1. US Public Law 91–596. The Occupational Safety and Health Act 1970
2. Gardner, A. Ward (Ed.), *Current Approaches to Occupational Medicine*, 205, John Wright & Sons, Bristol (1979)
3. Health and Safety Executive, *The Health and Safety (First Aid) Regulations 1981* (SI 1981 No. 917), HMSO, London (1981)
4. Health and Safety Executive. *First aid at work. The Health and Safety (first Aid) Regulations 1981. Approved Code of Practice*, Publication No. L74, HSE Books, Sudbury (1997)
5. Health and Safety Executive, *Basic advice on first aid at work*, Publication No. IND(G)215L, HSE Books, Sudbury (1997) (Single copies free)

Further reading

Croner's Handbook of Occupational Hygiene, loose-leaf publication with regular updates, Croner Publications Ltd, Kingston-upon-Thames
Gill, F.S. and Ashton, I., *Monitoring for Health Hazards at Work*, Blackwell Science, Oxford (1998)
Harrington, J.M., Gill, F.S., Aw, T.C. and Gardiner, K., *Occupational Health Pocket Consultant*, Blackwell Science, Oxford (1998)
Raffles, P.A.B., Adams, P., Baxter, P.S and Lee, W.R. (Eds), *Hunter's Diseases of Occupation*, 8th edn, Hodder and Stoughton, London (1992)
Rom, W.N. (Ed.), *Environmental and Occupational Medicine*, Little Brown & Co., Boston (1992)
Seaton, A. et al., *Practical Occupational Medicine*, Edward Arnold, London (1994)

Chapter 2

Occupational diseases

Dr A. R. L. Clark

2.1 Introduction

Large companies may employ a number of specialists in the field of
health, hygiene and safety who co-operate as a team and pool their
particular expertise, but in small firms the safety adviser is often the only
local source of advice on these matters, other expertise being brought in
when considered necessary.

The task of safeguarding the health of persons at work is a formidable
one especially during periods of rapid technological and organisational
change. Thousands of chemical substances are used by industry and in
commerce but only about 800 of those in common use have been
recognised as presenting a risk to the health of workpeople.

This chapter provides a brief introduction to some of the more
important diseases and materials that cause them. Conveniently, these fall
into four major areas covering illnesses and diseases due to:

chemical agents   (sections 2.2–2.10)
physical agents   (sections 2.11–2.15)
biological agents   (section 2.16)
psycho-social causes   (section 2.17).

In addition, section 2.18 deals with target organs.

2.2 Toxicology

There is no such state as absolute safety in the use of chemicals since all
chemicals are toxic to a degree depending on the dose. The toxicity of a
substance is its potential to cause harm on contact with body tissues.
Toxicology is the scientific study of the medical effects on living beings of
poisons[1,2].

To determine these effects, which may be acute or chronic, toxicity
testing may be carried out on animals, man (in vivo tests) and in test
tubes (in vitro tests). The tests are carried out in three stages.

Stage one – tests to establish acute effects

These tests aim to establish the lethal dose (LD) or lethal concentration (LC) of a substance and are carried out on mice, rats and sometimes larger animals. The degree of toxicity is indicated by the percentage of test animals that are killed by a single dose. Thus LD50 is the dose that will kill 50% of the test animals and is known as the median lethal dose. The values of LD20 and LD90 relate to doses that will kill 20% and 90% respectively. Toxicity units are specified as mg (or g) of poison/kg of body weight.

Typical degrees of LD50 toxicity[3] are:

extremely toxic	1 mg or less
highly toxic	1–5 mg
moderately toxic	50–500 mg
slightly toxic	0.5–5 g
practically non-toxic	5–15 g
relatively harmless	15 g or more

From these measures it is possible to calculate the toxic dose of a man of known weight.

Similarly, the lethal concentrations quoted for LC20, LC50 and LC90 are those that will kill 20%, 50% and 90% of test animals in a certain time (say 48 hours) when exposed in a chamber with a controlled atmosphere of a particular gas, fume, vapour, dust etc. Inhalation tests require the concentrations and exposure times to be recorded to ascertain the uptake by the animal.

In tests for irritant response, the substances can be applied in the eye (Draize test) or on the skin. Skin sensitivity patch tests may be undertaken on man.

Stage two – tests to determine chronic effects

The harmful effects are investigated over 90 days and involve giving regular sub-lethal doses of the substances by the expected human route of exposure, but not exceeding the likely human dose. The tests are carried out on rats and mice. Observations are made of behaviour, growth, food intake, urine, faeces, biochemistry of blood, electrolytes, urea, sugar, fat and the metabolites of the test substances. Teratology studies, using similar techniques, are carried out during the early stages of pregnancy.

Stage three – tests for carcinogenicity

These tests require a large number of test animals with known medical history and of pure strains and involve assessing the rate of tumour induction making allowance for possible spontaneous tumour occurrences.

Other tests include the Ames test in which a strain of bacteria, such as *Salmonella typhimurium*, is mixed with rat liver and the test substance then incubated for two days. The carcinogenicity is indicated by the number of mutants induced. This test is sensitive, quick and cheap. Also tissue cultures tests in which cells from a test animal are cultured in an isotonic fluid medium and the effects of adding the test substance are observed.

Each stage of toxicity testing takes about two years and at the end of it, as with all animal testing, there is difficulty in extrapolating the results to man especially when animals are tested in conditions to which man is not exposed.

2.2.1 Portals of entry

Occupational poisons gain entry to the body via the lungs, skin and sometimes the gut. Absorption of a poison depends on its physical state, particle size and solubility. Of the substances entering the lung some may be exhaled, coughed up and swallowed, attacked by scavenger cells and remain in the lung or enter the lymphatics. Soluble particles may be absorbed into the blood stream. The skin is protective unless abraded when soluble substances can penetrate to the dermis, as they may also do via the hair follicles, sweat and sebaceous glands, and then be absorbed into the blood stream.

In pregnancy, harmful substances in the mother's body may cross the placenta to affect the unborn baby.

2.2.2 Effects

Effects may be acute, i.e. of rapid onset and short duration; or chronic, i.e. of gradual onset and prolonged. They may be local, occurring at the site of contact only, or general following absorption. Toxic substances may disturb normal cell function, damage cell membranes, interfere with enzyme and immune systems, RNA and DNA activity. Pathological response may be irritant, corrosive, toxic, fibrotic, allergic, asphyxiant, narcotic, anaesthetic and neoplastic.

2.2.3 Metabolism

Most substances absorbed will be carried by the blood stream to the liver where they may be rendered less harmful by a change in their chemical composition. However, some may be made more toxic, e.g. naphthylamine which is responsible for bladder cancer and tetra-ethyl lead which is converted into the tri-ethyl form and is toxic to the central nervous system.

2.2.4 Excretion

The body eliminates harmful substances in the urine, lungs and less commonly the skin. Some are also excreted in the faeces and milk. The time taken to reduce the concentration of a substance in the blood by 50% is known as its biological half-life. Similarly, the time for a 50% fall in concentration of a substance or its metabolite in urine or breath, after cessation of exposure, is its half-life in that medium. This is important in the design of screening tests[4].

2.2.5 Factors influencing toxicity

A number of factors are important when considering the toxic effect of a substance on the body. These include:

1 The inherent potential of a substance to cause harm.
2 Its ease of body contact and entry: work method, particle size and solubility.
3 Dose received (concentration and time of exposure).
4 Metabolism in the body (bio-transformation) and its half-life.
5 Susceptibility of the individual which depends on a number of factors:
   (a) Body weight; the same dose of a substance is more damaging to the smaller person.
   (b) The extremes of age in the working population are more prone to skin damage.
   (c) Fair skinned persons are more liable than the dark skinned to chemically induced dermatitis, and to radiation induced skin cancer.
   (d) Physical and physiological differences between the sexes may cause a variation in toxic response.
   (e) Immunological, nutritional and genetic defects.
   (f) Failure to reach a set health standard for work may expose the individual to greater risk.
   (g) Inadequate level of training, information, supervision and protection.

2.2.6 A no-adverse-effect level

In the setting of Occupational Exposure Standards (OES)[5], which are health based exposure concentration limits, it is necessary to establish, with reasonable certainty, the airborne concentrations which will not result in ill-health even if inhaled day after day. This concentration may be derived from the no-adverse-effect level in animal species which is arrived at after careful epidemiological and toxicological tests. This level relates to the average workperson but will not apply to individuals who, under certain circumstances, are susceptible and, therefore, at special risk[6].

2.2.7 Epidemiology

Epidemiology may be defined as the study and distribution of disease in human populations. The need for a study may be triggered by suspicions about an individual case, a complaint or the occurrence of a cluster of cases.

An initial *descriptive study* of a cross-section of the affected population is undertaken to determine:

What is the disease	
Who is affected	(sex, age, race, social class and occupation)
Where does it occur	(factory, workshop, room, laboratory, area of site)
When does it occur	(time, day, shift)
How are persons infected	(skin contact, airborne, body fluids)

From this study a hypothesis may be formed which needs to be tested by an analytical investigation involving a case-control study and a cohort study.

A *case-control study* compares persons who have the disease with those without to establish whether the suspect cause occurs more frequently among those with than those without the disease.

A *cohort study* compares those exposed to the suspect cause with those who are not to determine if more persons exposed to the cause develop the disease than those who are not.

An *incidence study* is essential for determining the risk, magnitude and causative factors of a disease:

$$\text{the incidence rate} = \frac{\text{no. of new cases of the disease}}{\text{population at risk}} \text{ over a period of time}$$

2.3 Disease of the skin

2.3.1 Non-infective dermatitis

The term 'dermatitis' simply means an inflammation of the skin. When the condition is due to contact with a substance at work it is called 'occupational' or 'industrial' dermatitis. It is a common cause of occupational disease but the number of cases is declining owing to improved work conditions.

The skin has two layers, the outer layer is called the 'epidermis' and the inner the 'dermis'. The epidermis has a protective function. It consists of densely packed flat cells, thicker in some areas, like the palms of the hands, which are more subject to injury. It is covered by a moist film known as an 'acid mantle', made up of secretions from sweat and sebaceous glands, that helps to protect from acids, alkalis, excessive water, heat and friction by preventing the skin from drying out. The natural grease of the skin can be removed by solvents. In the deeper layer of the epidermis are pigment cells which produce the 'tan' following exposure to sunlight and protect the body from ultraviolet radiation.

Some persons are more susceptible to skin damage than others, particularly the young, those with soft, sweaty skin, the fair complexioned and those with poor personal hygiene. Occupational dermatitis can affect any part of the body, but the hands, wrists and forearms are most commonly involved. Damage to the skin may follow exposure to chemical and biological substances as well as physical agents. Dermatitis is of two kinds: irritative and sensitising – the former is four times more common. Chemicals which cause irritative dermatitis include acids, alkalis, cement, solvents, some metals and their salts. Their effect on the skin depends on the concentration and duration of exposure, and will affect most people in contact with them. At first the response may be minor, but it worsens with repeated contact.

Sensitisers, on the other hand, do not cause dermatitis until the individual has first become sensitised by them. This involves an allergic response in the tissues initially, dermatitis follows on subsequent exposure. Once sensitisation has occurred a small dose may be sufficient to cause a rash. Sensitisers include chrome-salts, nickel, cobalt, plastics made of epoxy, formaldehyde, urea or phenolic resins, rubber additives, some woods and plants[7]. Some substances act as both irritant and sensitiser, e.g. chrome, nickel, turpentine and mercury compounds.

2.3.1.1 Symptoms

The onset of dermatitis may be unnoticed, especially as it usually clears up when away from work, i.e. at weekends and holidays. On return to work and further exposure the condition recurs, worsening with each subsequent contact. The skin, at first rough and raw, may itch, become cracked and sore, prompting the individual to seek medical advice. The rash may be diffuse, as with eczema, or pimple-like as with acne – the former following exposure to irritative and sensitising agents, and the latter from exposure to mineral oils, pitch and chlorinated hydrocarbons. Patch testing, in which a dilute quantity of chemical is applied to the skin under a plaster and left for several hours to see if a reaction develops, is useful only for determining allergic response to chemicals but requires specialist interpretation.

2.3.1.2 Protective measures

Persons with dermatitis or sensitive skin may need to be excluded from certain kinds of work. Good personal hygiene is essential and barrier creams may be helpful. Protective clothing should be considered.

2.3.2 Cancer of the scrotum

The first occupational skin cancer was reported by Percivall Pott in 1775, among chimney sweeps. In those days children were apprenticed to master sweeps to climb inside and clean chimneys; their skin became ingrained and their clothes impregnated with soot, and as they seldom washed or changed their clothes the skin was constantly irritated. From puberty onwards a 'soot wart' might appear on the scrotum and develop into a cancer. In 1820, Dr Paris wrote of the influence of arsenical fumes affecting those engaged in copper smelting in Cornwall and Wales, giving rise to a cancerous disease of the scrotum similar to that affecting sweeps[8]. From 1870 a number of substances in a variety of industries were found to cause scrotal cancer – shale oil in those engaged in oil refining and cotton mule spinning; pitch and tar in those making briquettes from pitch-containing coal dust; mineral oil used by engineers and gunsmiths, and paraffin in refinery workers. Others at risk include creosote-timber picklers and anthracene chemical workers, and also sheep-dippers using arsenic[9].

Workers' clothes become begrimed with the offending substance, making close contact with the scrotum, the wrinkled skin of which favours the harbouring of the carcinogen.

The cancer begins as a wart, which enlarges and hardens, then breaks down into an ulcer with spread of malignant cells to neighbouring glands and other parts of the body.

Skin cancer is often due to polycyclic aromatic hydrocarbon of the benzpyrene or benzanthracene type. It has also been attributed to sunlight, ionising radiation and arsenic compounds.

2.3.2.1 Prevention

The use of non-carcinogen oils: carcinogens can be removed from mineral oil by washing with sulphuric acid or solvents. Workers should be educated to avoid contact as much as possible. The use of splash guards on machinery, protective clothing, avoidance of an oily rag placed in a pocket, which could spread oil through the clothes to the scrotum. To wash the hands before toilet and to have a daily bath. Clothes should be kept reasonably clean and a laundry service provided, so that overalls can be changed once or twice a week as need requires. Workers should not wear their dirty overalls after duty, but be encouraged to change into their home clothes. Workers also should be medically examined prior to employment and periodically to ensure that their skin is clear, and be encouraged to report to the doctor any doubtful 'wart' that might appear.

2.3.3 Coal tar and pitch

The destructive distillation of coal yields a variety of products, depending on the temperature at which distillation takes place, e.g.

	Temp°C	Product
GAS ↑	200	light oil
Ammoniacal liquid ← COAL → TAR ←	250	carbolic substances
↓ ↓		
COKE Residue = pitch	300	creosote
	350	anthracene

Distillation at high temperatures results in aromatic polycyclic hydro-carbons retained in the pitch which are harmful to health. Pitch is used in many industries: briquetting of coal, roofing materials, waterproofing of wood, manufacture of electrodes, impermeable paper, optical lenses, dyestuffs and paints.

2.3.3.1 Symptoms

Exposure of a worker to pitch dust or vapour may harm the skin by causing irritation, tumour or dermatitis. Irritation is the earliest and commonest reaction, occurring after a few days or weeks of exposure and affecting the face and neck. There is complaint of itching or burning, aggravated by cold, wind or sunlight (Pitch Smarts). Usually it clears up soon after exposure ceases. Benign tumours or warts occur on exposed areas of skin, chiefly the face, eyelids, behind the ears, the neck, arms and, occasionally, on the scrotum and thighs. Their recurrence is related to duration and degree of exposure to pitch[10]. Many regress spontaneously, especially those appearing early, but some undergo malignant change, particularly those appearing in the older age groups. They need to be removed and examined under the microscope, i.e. biopsied, to check for any malignant change. A variety of other skin conditions may occur such as darkening and thickening of the skin, acne, blackheads, cysts and boils, pitch burns and scarring. There is also a risk of damage to the cornea.

2.3.3.2 Prevention

Pitch dust and vapour must be avoided by transporting the raw material in a liquid or granular state and enclosing the process as far as possible. Workers require clean protective clothing for head, neck and forearms and eye protection should be worn[11]. Employees ought to be warned of the risk, and advised to report any skin disease which develops. Good personal hygiene is essential, and adequate wash and shower facilities need to be provided. Barrier creams applied before work are helpful. Those susceptible to warts should be excluded from further exposure, and each worker needs to be medically examined regularly to detect possible skin disorders.

2.4 Diseases of the respiratory system

2.4.1 Pneumoconiosis

The term pneumoconiosis means 'dust in the lung', but medically refers to the reaction of the lung to the presence of dust[12].

2.4.1.1 Body defence to inhalation of dust

During inspiration particles of dust in the air larger than $10\,\mu$m in diameter are filtered off by the nasal hairs. Others, which enter through the mouth, are deposited in the upper respiratory tract. Particles between 5 μm and 10 μm tend to settle in the mucus covering the bronchi and bronchioles and are then wafted upward by tiny hairs (ciliary escalator) towards the throat. They are then coughed or spat out, though some may be swallowed. Particles less than 5 μm in diameter are more likely to reach the lung tissue. However, fibres (e.g. asbestos) which predispose to disease have a length to diameter ratio of at least 3:1 with a diameter of 3 μm or less; the longer the fibre the more damaging it may be.

2.4.1.2 Respirable dust

Respirable dust is that dust in the air which on inhalation may be retained by the lungs. The amount of dust retained depends on the duration of exposure, the concentration of dust in the respired air, the volume of air inhaled per minute and the nature of the breathing. Slow, deep respirations are likely to deposit more dust than rapid, shallow breathing. Dust in the lung causes a tissue reaction, which varies in nature and site according to the type of dust. Coal and silica dust involve the upper lungs whereas asbestos involves the lower lungs.

2.4.1.3 Causes of pneumoconiosis

(a) *Benign* The inhalation of some metal dusts, such as iron, tin and barium, results in very little structural change in the lungs and, therefore, few symptoms. The tissue reaction, nevertheless, is detectable on X-ray as a profusion of tiny opacities.
(b) *Symptomatic* The most important causes include coal dust, silica and asbestos. Symptoms of cough and breathlessness develop usually after many years of exposure, but only in the later stages of disease.

Beryllium dust causes acute and chronic symptoms. Early features are breathlessness, cough with bloody sputum and chest pain. Recovery follows removal from exposure, but a chronic state can develop insidiously with cough, breathlessness and loss of weight.

Organic dusts, such as mouldy hay, when inhaled cause a disease known as extrinsic allergic alveolitis with 'flu-like symptoms; cough and difficulty in breathing occur within a few hours of exposure. Repeated exposure leads to further lung damage and chronic breathlessness.

Talc is a white powder consisting of hydrous magnesium silicate. Although some talc presents little risk to health, commercial grades may contain asbestos and quartz and provoke pneumoconiosis and lung cancer.

Cobalt combined with tungsten carbide forms a hard metal used for the cutting tips of machine tools and drills. Inhalation of the dust may give rise to fibrosis of the lungs causing cough, wheezing and shortness of breath.

Man-made mineral fibres irritate the skin, eyes and upper respiratory tract. A maximum exposure limit has been set based on the risk of lung cancer because a 'no-adverse-effect' level cannot be established with reasonable certainty[6].

2.4.1.4 Diagnosis of pneumoconiosis

This depends on:

1 A complete occupational history of all jobs.
2 A characteristic appearance on the chest X-ray. There is an international grading system which is used to assess radiologically the extent of the disease.
3 A clinical examination.
4 Lung function tests.
5 In some cases involving organic dust, specific blood tests.

2.4.2 Silicosis

Silicosis: the commonest form of pneumoconiosis is due to the inhalation of free silica.

Free silica (SiO_2) or crystalline silica occurs in three common forms in industry: quartz, tridymite and cristobalite. A cryptocrystalline variety occurs in which the 'free silica' is bound to an amorphous silica (non-crystalline). It includes tripolite, flint and chert. Diatomite is the most common form of amorphous silica capable of producing lung disease. Some of these forms can be altered by heat to the more dangerous crystalline varieties, such as tridymite and cristobalite. e.g.

Quartz

Cryptocrystalline $\Big\}$ $\to$ —→ tridymite $\to$ $\to$ cristobalite $800°C^+$ ————→

Amorphous

2.4.2.1 Lung reaction

Industrial exposure occurs in mining, quarrying, stone cutting, sand blasting, some foundries, boiler scaling, in the manufacture of glass and ceramics and, for diatomite, in the manufacture of fluid filters. Particles of free silica less than 5 μm in diameter when inhaled are likely to enter the lungs and there become engulfed by scavenging cells (macrophages) in the walls of the tiniest bronchioles. The macrophages themselves are destroyed and liberate a fluid causing a localised fibrous nodule which obliterates the air sacs. The nodules are scattered mainly in the upper halves of the lungs. They gradually enlarge to form a compressed mass of nodules. Sometimes a single large mass of tissue may occur, known as progressive massive fibrosis. If much of the lung is affected the remaining healthy tissue is likely to become over-distended during inhalation.

2.4.2.2 Symptoms

There are no symptoms in the early stage. Later the initial complaint is of a dry morning cough. Next occurs some breathlessness, at first noticeable on exercise but, as destruction of lung tissue proceeds, breathlessness worsens until it is present at rest. The interval between exposure and the onset of symptoms varies from a few months in some susceptible individuals to, more usually, many years, depending on the concentration of respirable free silica and the exposure time at work. Silicosis is the one form of pneumoconiosis which predisposes to tuberculosis, when additional symptoms of fever, loss of weight and bloody sputum may occur. In the presence of gross lung destruction the blood circulation from the heart to the lung may be embarrassed and result in heart failure.

2.4.2.3 Diagnosis

This depends on a history of exposure and, in the early stages, a chest X-ray showing tiny radio opaque nodules and, later, a history of cough and breathlessness and sounds in the chest detectable with a stethoscope. Lung function tests may be helpful, but usually not until the late stages.

2.4.2.4 Medical surveillance

Where exposure to free silica is a recognised hazard, a pre-employment medical is advised, which should enquire into previous history of dust exposure, of respiratory symptoms, with examination of the chest, lung function testing and a chest X-ray. The medical should be repeated periodically as circumstances demand.

2.4.2.5 Prevention

Reduction of the dust to the lowest level practicable and where necessary by the provision of personal respiratory protective equipment.

2.4.3 Asbestosis

There are three important types of asbestos, blue (crocidolite), brown (amosite) and white (chrysotile). Asbestosis is a reaction of the lung to the presence of asbestos fibres which, having reached the bronchioles and air sacs, cause a fibrous thickening in a network distribution, mainly in the lower parts of the lung[13]. There follows a loss of elasticity in the lung tissue (relative to the concentration of fibres inhaled and the duration of exposure), resulting in breathing difficulty.

Among those at risk are persons engaged in milling the ore, the manufacture of asbestos products, lagging, asbestos spraying, building, demolition, and laundering of asbestos workers' overalls.

Symptoms develop slowly after a period of exposure which varies from a few to many years. In some cases exposure may have begun so long ago that it cannot be recalled. Breathlessness occurs first and progresses as the lung loses its elasticity. There may be little or no cough and chest pain seldom occurs. The individual becomes weak and distressed on effort and, eventually, even at rest. Unless periodic medicals are introduced the diagnosis will not be made until symptoms appear. Early diagnosis is essential in order to prevent further exposure and an exacerbation of the condition. Asbestosis predisposes to cancer of the bronchus, a risk increased by cigarette smoking. The chest should be X-rayed every two years and special lung function tests are helpful. Diagnosis depends on history of exposure, chest X-ray, lung function testing, symptoms and physical signs.

2.4.4 Mesothelioma

Mesothelioma is a malignant tumour of the lining of the lung (pleura) or abdomen (peritoneum). The abdominal form is less common. The disease is significantly related to exposure to asbestos, especially the blue and brown varieties. However, in some 10–15% of cases there is no such history of exposure[13]. Those at risk are miners, manufacturers of asbestos, builders and demolition workers, and even residents in the neighbourhood of blue asbestos working. While the exposure time may have been minimal, there is no safe threshold of dose below which there is no risk of asbestos-related disease. The onset of the disease is delayed by some 20 to 50 years.

2.4.4.1 Symptoms

The lung variety of tumour is more common. Symptoms begin with a gradual onset of breathlessness, particularly noticeable on effort, and due to the growth of tumour and fluid compressing the lung. There may occur pain on one side of the chest, with tenderness, cough and fever. More obvious is a rapid loss of weight and weakness. A chest X-ray reveals an opacity on one side of the chest suggestive of the tumour. The symptoms of the abdominal form also develop slowly, beginning with a swelling, loss of weight, impaired appetite and weakness. Death usually follows within two years of making the diagnosis.

2.4.5 Other dust causes of lung cancer

These include: chromate, in the manufacture of chromate from the ore; nickel compounds in the refining of nickel; benzpyrenes in coke-oven work; uranium and radon; and arsenic compounds in mining.

2.4.6 Bronchial asthma

Bronchial asthma is defined as breathlessness due to narrowing of the small airways and it is reversible, either spontaneously or as a result of treatment. It may follow inhalation of a respiratory sensitiser or an irritant toxic substance. Symptoms due to sensitisation may be delayed for weeks, months or even years; symptoms due to a toxic substance occur within hours of inhalation, resolve spontaneously but can persist indefinitely. The toxic response is called *reactive airways dysfunction* (RAD) syndrome. Most cases of occupational asthma are due to sensitisation and are listed[14] as prescribed diseases for purposes of statutory compensation. The sensitising substances listed are:

1 Isocyanates.
2 Platinum salts.
3 Epoxy resin curing agents.
4 Colophony fumes.
5 Proteolytic enzymes.
6 Animals and insects in laboratories.
7 Flour and grain dust.
8 Antibiotic manufacture.
9 Cimetidine used in manufacturing cimetidine tablets.
10 Hard wood dusts of cedar, oak and mahogany.
11 Ispaghula used in the manufacture of laxatives.
12 Caster bean dust.
13 Ipecacuanha used in the manufacture of tablets.
14 Azodicarbonamide used in plastics.
15 Glutaraldehyde, a cold disinfectant used in the health service.
16 Persulphate salts or henna used in hair dressing.
17 Crustaceans or fish products used in the food processing industry.
18 Reactive dyes.
19 Soya bean.
20 Tea dust.
21 Green coffee bean dust.
22 Fumes from stainless steel welding.
23 Any other sensitising agent inhaled at work.

Respiratory sensitisers may be referred to as *asthmagens*. In 1989 Surveillance of Work Related Respiratory Disease (SWORD) was started and contains reports by respiratory and occupational physicians.

Other asthma-like diseases are found.

Byssinosis occurs in workers in the cotton processing industry who may develop tightness of the chest on Mondays which decreases as the week progresses. However, with continuing exposure to cotton dust they are affected for more days of the week. Steam treatment of the raw cotton can prevent chest symptoms from this material.

An allergic lung reaction also occurs after exposure to spores on sugar cane (*bagassosis*). The sugar cane spores can be killed by spraying with propionic acid.

2.4.7 Extrinsic allergic alveolitis (farmer's lung)

A disorder due to inhalation of organic dust and characterised by chest tightness, fever and the presence of specific antibodies in the blood. Typical examples are:

Disease	Exposure	Allergen
Farmer's lung	mouldy hay	mould
Malt worker's lung	mouldy barley	mould
Bagassosis	mouldy sugar cane	mould
Bird fancier's lung	bird droppings	protein
Animal handler's lung	rats' urine	protein

2.5 Diseases from metals

2.5.1 Lead

Lead (Pb) is a relatively common metal, mined chiefly as the sulphide (galena) in many countries – USA, Australia, USSR, Canada and Mexico[15]. In this country we use about 330 000 tonnes of lead annually, much of which comes from recycled scrap.

Lead has a great variety of uses, e.g. (percentages approximated from annual production figures issued by World Bureau of Metal Statistics, London):

Electric batteries	27%
Electric cables	17%
Sheet, pipe, tubes	16%
Anti-knock in petrol	11%
Solder and alloys	9%
Pottery, plastics, glass, paint	4%
Miscellaneous	15%

Lead, as a fume or dust hazard, is therefore met in many industries. The pure metal melts at 327°C and begins to fume at 500°C, but the presence of impurities alters these properties and may form a slag on its surface and thereby reduce fuming, except at higher temperatures. Particle size and solubility are important factors governing the absorption of lead via the lungs. In the gut, however, solubility differences of ingested compounds are of less significance. Among lead miners lead poisoning does not occur due to the insolubility of the sulphide ore.

2.5.1.1 Inorganic lead

Inorganic lead can enter the body by inhalation or ingestion[16]. Up to about 50% of that inhaled is absorbed and only about 10% of that ingested. It is then transported in the blood stream and deposited in all tissues, but about 90% of it is stored in the bone. It is a cumulative poison; excretion is slow and occurs mainly in the urine and faeces.

Symptoms Early features are vague and include fatigue, loss of appetite, and metallic taste in the mouth. Constipation is the commonest complaint and is sometimes associated with abdominal pain. This may be so severe as to mimic an acute abdominal emergency. Classically, a blue line appears along the margin between the teeth and gums, but this usually occurs only in the presence of infected teeth and is indicative of lead exposure rather than poisoning.

Lead interferes with the normal formation of haemoglobin, causing anaemia, but the diagnosis of excessive absorption should be made before anaemia appears. The same interfering mechanism causes abnormal products to appear in the urine, e.g. amino laevulinic acid (ALA) which is a useful indicator of excessive lead absorption or poisoning.

Paralysis, though rare nowadays, can occur as wrist or foot drop due to the effect of lead on nerve conduction. It may begin with a weakness in the fingers and wrists, which is a useful early sign.

Lead is transported in the blood and can cross the placental barrier in pregnant women and affect the unborn child. Abortion was common in women employed in lead industries during the nineteenth century and was believed to be due to excessive lead absorption. The brain can also be affected, a condition known as encephalopathy, causing abnormal behaviour, convulsions, coma and death. Children are much more susceptible than adults.

Because of the excretion of lead in the urine, kidney damage is a likely long-term effect.

2.5.1.2 Organic lead

Tetra-ethyl and tetra-methyl lead are the most important organic forms used in industry, especially in petrol to improve the octane rating. These substances can be absorbed via the lungs and the skin. In the liver they are changed respectively to tri-ethyl and tri-methyl lead, which are much more toxic. They have a particular predilection for the brain and cause psychiatric disturbance, headache, vomiting, dizziness, mania and coma. Excretion occurs mainly via the urine. The blood is less affected than with inorganic lead.

2.5.1.3 Biological monitoring

For lead workers periodic medical examination is a statutory requirement. Blood samples should be taken as required for haemoglobin and lead. Lead level in normal blood is about $20 \mu g/100 ml$ but for lead workers can be $40–60 \mu g/100 ml$. The acceptable upper limit of blood lead concentration in adults is $60 \mu g/100 ml$ except men who have worked in lead for many years. For young persons is $50 \mu g/100 ml$ and for women of child-bearing age the limit is $30 \mu g/100 ml$.

A useful indicator of excessive lead effect is the presence of zinc protoporphyrin (ZPP). It can be measured from a small quantity of blood obtained by finger-prick. For confirmatory evidence of excessive lead absorption or poisoning, urine estimation of amino laevulinic acid is helpful. Inorganic lead is best monitored by blood sampling and organic lead by urine sampling.

2.5.2 Mercury

Mercury (Hg) occurs naturally as the sulphide in the ore known as cinnabar, and also in the metallic form quicksilver. It is mined chiefly in Spain, but also in Italy, Russia, USA and elsewhere. The ore is not particularly hazardous to miners, as the sulphide is insoluble. Risk is greater in other industries, such as in the manufacture of sodium hydroxide and chlorine, electrical and scientific instruments, fungicides, explosives, paints and in dentistry.

2.5.2.1 Symptoms

Acute mercury poisoning is rare but can occur following the inhalation of quicksilver – it being very volatile at room temperature. There is particular risk should spillage occur in an enclosed space. About 80% of that inhaled can be absorbed[17], and a few hours later there occurs cough, tight chest, breathlessness and fever. Symptoms last a week or so, dependent upon degree of exposure, but its effects are reversible. Acute poisoning may also occur by ingestion of soluble salts, such as mercuric chloride which has a corrosive action on the bowel, causing bloody diarrhoea.

Ingestion of metallic mercury is not generally toxic as it is not absorbed.

Chronic poisoning is the more usual presentation, following absorption by lung or gut of soluble mercury salts. Symptoms develop almost imperceptibly, usually beginning with a metallic taste in the mouth and sore gums. Later tremor of the hands and facial muscles develops; gums may bleed and teeth loosen. Personality changes of shyness and anxiety, inability to concentrate, impaired memory, depression and hallucinations may occur. As excretion is mainly via the urine, the kidney is subject to damage.

Organic mercury can be absorbed via the lung, gut and skin, and also cause chronic poisoning. There are two varieties: aryl and alkyl, and they have different effects on the body. The aryl variety, of which phenyl mercury is an example, has a similar metabolic pathway to inorganic mercury and has a similar clinical effect.

Alkyl mercury is much more dangerous – methyl mercury is an example. It causes irreversible damage to the brain, resulting in a constriction of visual fields, disturbance of speech, deafness and inco-ordination of movement. Most of it (90%) is excreted without change, slowly in the faeces.

All forms of mercury may give rise to dermatitis. Mercury can cross the placental barrier and affect the unborn child of exposed mothers.

2.5.2.2 Health surveillance

Those at risk should be medically examined periodically and attention paid to the mouth, tremor of the hands (a writing test is useful), personality, and for those exposed to methyl mercury, vision, hearing and co-ordination. The urine should be checked for protein and mercury excretion. Mercury does not normally occur in urine, but may be detected in some persons with no apparent occupational exposure. In organic exposure, owing to the different metabolic pathway from that of inorganic, the urine concentration does not correlate with body levels. The upper limit which requires further investigation is for inorganic mercury 1000 nmol/litre and, for organic mercury, 150 nmol/litre[18].

2.5.3 Metal fume fever

Inhalation of the fume of some metal oxides such as zinc, copper, iron, magnesium and cadmium causes an influenza-like disease. Similar effects may follow the inhalation of polytetrafluoroethylene (ptfe) fumes. Usually there is recovery within one or two days. Zinc fume fever is probably a very common disease, the diagnosis of which is often missed because of the short duration of the illness. Cadmium fume inhalation can be much more serious. It has a half-life of several months (see section 2.5.9).

2.5.4 Chromium

Chromium (Cr) is a silvery hard metal used in alloys and refractories. Chrome salts are used in dyeing, photography, pigment manufacture and cements. Electroplating tanks contain solutions of chromic acid which forms a mist during the electrolysis process.

Chromates and dichromates used in cement manufacture and chromium plating may cause skin irritation or ulceration and chrome ulcers in the skin of the hands or in the inside of the nose where the ulcer may penetrate the cartilage of the nasal septum.

2.5.5 Arsenic (As)

Inorganic arsenic compounds cause irritation of the skin and may produce skin cancer. It is used in alloys to increase hardness of metals, especially with copper and lead.

2.5.6 Arsine (arseniuretted hydrogen – AsH_3)

Arsine is a gas which arises accidentally in many metal working industries. It damages the red blood cells, releasing the red pigment haemoglobin from them. This may cause jaundice, anaemia and the urine may appear red due to the presence of haemoglobin pigment. Poisoning by arsine can result in rapid death. Organic arsenic compounds have been used as war gases, and can produce severe and immediate blistering of the skin and severe lung irritation (pulmonary oedema).

2.5.7 Manganese (Mn) and compounds

This is used to make manganese alloy steels, dry batteries and potassium permanganate which is an oxidising agent and a disinfectant. Poisoning is rare and follows inhalation of the dust causing acute irritation of the lungs and affects the brain leading to impaired control of the limbs rather like Parkinson's disease.

2.5.8 Nickel (Ni) and nickel carbonyl (Ni(Co)₄)

Nickel is a hard blue-white metal used in electroplating and in a range of alloys. Nickel salts (green) cause skin sensitivity (nickel itch). Nickel carbonyl (a colourless gas) causes headache, vomiting and later pulmonary oedema.

2.5.9 Cadmium (Cd)

This metal is used in alloys, rust prevention, solders and pigments. A fume may be released during smelting, alloy manufacture or when rust-proofed metals are heated, e.g. in welding cadmium-plated metals, which produces irritation of the eyes, nose and throat. With continued exposure tightness of the chest, shortness of breath and coughing may increase and can lead to more severe lung damage which may be fatal.

Long-term damage by smaller quantities of dust or fumes may lead to loss of elasticity of the lungs. Cadmium may cause kidney damage and while it has been suggested that lung cancer may occur after cadmium exposure this has not been proved in man.

2.5.10 Vanadium (V)

This material occurs as vanadium ore and is found in petroleum oil. It is also used to make alloy steels and as a catalyst in many chemical reactions. Exposure to the metal occurs when oil-fired boilers are cleaned and manifests itself in eye irritation, shortness of breath, chest pain and cough. The tongue becomes greenish-black in colour. Severe cases may develop broncho-pneumonia. Removal from contact with the dust usually leads to rapid recovery.

2.6 Pesticides

2.6.1 Insecticides

Various organo-phosphorus compounds are used; two of the commonest are demeton-S-methyl and chlorpyrifos. Poisoning causes headaches, nausea and blurred vision. Further symptoms include muscle twitching, cramps in the belly muscles, severe sweating and respiratory difficulties. Extreme exposure may lead to death. All these effects are due to interference with a chemical enzyme called cholinesterase which is concerned with the passage of nerve impulses. The level of this enzyme in the worker's blood can be measured and if it falls below a certain value the worker must be removed from contact with the chemical until his blood returns to normal. The appropriate protective clothing must be worn at all times when working with these materials.

2.6.2 Herbicides

Commonly used as a weedkiller (e.g. paraquat). Ingestion may result in damage to the liver, kidneys and lung. There is no antidote and death occurs in about half the cases.

2.7 Solvents

A solvent is a liquid that has the power to dissolve a substance: water is a common example[19]. In industry organic liquids are often used as solvents, and these are mainly hydrocarbons used as degreasing agents and in the manufacture of paints and plastics.

Examples of solvents (classification after Matheson[20])

Hydrocarbons	
(i) Aromatic	Benzene; toluene; styrene
(ii) Aliphatic	Paraffin; white spirit
Aliphatic alcohols	Methyl alcohol; ethyl alcohol
Aliphatic ketones	Methyl-ethyl-ketone
Aliphatic ethers	Diethyl ether
Aliphatic esters	Ethyl acetate
Aliphatic chlorinated	Trichloroethylene; carbon tetrachloride
Non-hydrocarbons	Carbon disulphide

2.7.1 General properties

All organic solvents are volatile and have a vapour density greater than one, i.e. their vapours are heavier than air and will therefore settle at floor level; this is important to note when considering ventilation. With the exception of the chlorinated hydrocarbons they tend to be flammable and explosive and in the liquid form most have specific gravities of less than one so will float on water. In the event of a fire, attempt should not be made to extinguish with water, as the solvent will float away and the fire will spread. The chlorinated solvents, being neither flammable nor explosive but heavier than water, have been used as fire extinguishants.

2.7.2 Toxic effects

Solvents vary widely in their toxicological properties. In common they cause dermatitis by removing the natural grease from the skin, and narcosis by acting on the central nervous system; additionally some can damage the peripheral nerves, the liver and kidneys and interfere with blood formation and cardiac rhythm. Chlorinated solvents can decompose if exposed to a naked flame to produce acidic fumes (hydrochloric acid and small amounts of phosgene) which are harmful to the lungs. Any harmful effect is related to the amount of solvent absorbed.

Skin penetration varies with the solvent, hence in the list of Occupational Exposure Limits[5] some are designated 'skin', but other factors include surface area exposed and the thickness of the skin, e.g. less may be absorbed via the palms than the forearms while the scrotal area is most absorptive[21].

Absorption is also related to the breathing pattern, activity, obesity and addiction. Because of this individual variation, the amount taken up by the body is a more important estimate of potential harm than the concentration to which the body is exposed. Body uptake correlates well with blood concentration and to a less extent with quantities excreted in the urine[22].

However, periodic urine testing of excreted solvent or its metabolite is a more convenient means of biological monitoring[4]. The biological half-life of solvents is only a few hours. The half-life of some solvents is so short that biological monitoring of urine is not suitable, instead a metabolite must be used, such as mandelic acid for styrene and methyl hippuric acid for xylene.

2.7.3 Trichloroethylene

Structural formula:

Other names: Tri, Trike, Trilene.
Properties: Non-flammable
               Vapour density 4.54
               Specific gravity 1.45
               Boiling point 87°C
               MEL 100 ppm 8 hour TWA (skin)

Exposure to naked flames or red-hot surfaces can cause it to dissociate into hydrogen chloride, possibly with small amounts of phosgene or chlorine.

Use Its main use is as a solvent especially in the degreasing of metals. It has also been used as an anaesthetic. *Figure 2.1* shows a single compartment vapour type plant used for cleaning by solvents.

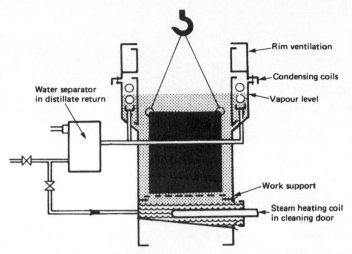

Figure 2.1 Cleaning by solvents: single compartment vapour type plant. (Courtesy ICI, PLC, Mond Division)

Metabolism Its main route into the body is via the lungs, where it is rapidly absorbed. Some is excreted into the expired air, while the remainder is converted to trichloroacetic acid and passed in the urine. It is usually cleared quickly from the body, but daily exposure may tend to its cumulation. The estimation of trichloroacetic acid in the urine is a useful test for checking excessive exposure and its concentration should not exceed 100 mg/litre urine, standardised to a specific gravity of 1.016. Samples should be collected at the end of a shift towards the end of a working week.

2.7.3.1 Harmful effects

Acute Trichloroethylene is a powerful anaesthetic and can be dangerous in confined spaces. Early features include headache, dizziness, and lack of concentration and eventually unconsciousness. Its vapour may cause irritation of the eye and the skin can be blistered by the liquid.

Chronic The main problem from repeated exposure is a dermatitis of the hands, due to the solvent's action in removing the normal grease of the skin which then becomes rough, red, raw, and cracks – a condition known as eczema. Some people become addicted to trichloroethylene, usually by repeated 'sniffing' of the vapour, or even drinking the fluid, and then display abnormal behaviour known as 'tri-mania'. Rare cases of sudden cardiac arrest have been reported in situations of gross short-term overexposure. After long-term exposure there have been a number of individual case reports of liver damage, and recently, following animal tests in the USA, it has been under suspicion as a carcinogen.

2.7.3.2 Prevention

Employees should be made aware of the risks. Local exhaust ventilation around the lips of vapour degreasing tanks is necessary, and in confined spaces good general ventilation is essential. In work areas, atmospheric monitoring is recommended to ensure that exposure is kept to a minimum and certainly below the Maximum Exposure Limit (MEL) of 100 ppm 8 hour TWA. Body absorption can be monitored by a urine sample taken at the end of a shift near the end of a working week and analysed for trichloroacetic acid. Those being tested must refrain from drinking alcohol as it inhibits excretion.

2.7.4 Carbon tetrachloride

Structural formula:

CI   CI
  \\   /
   \\ /
    C
   / \\
  /   \\
CI   CI

> *Properties*: Non-flammable
> Vapour density 1.5
> Specific gravity 1.6
> Boiling point 76.8°C
> OES-TWA 2 ppm (skin)

Use Its main use is in the manufacture of chlorofluorocarbons, also aerosols and refrigerants. It has been used in fire extinguishers and grain fumigation. Its use in dry cleaning has declined because of its toxicity.

Metabolism Carbon tetrachloride is absorbed into the blood mainly via the lungs, but also via the skin and gut. Some is excreted in the expired air and the remainder in the urine, but in altered form.

2.7.4.1 Harmful effects

Acute In common with other solvents it has a narcotic effect, with features varying from headache and drowsiness to coma and death. If taken by mouth it can cause abdominal pain, diarrhoea and vomiting. Acute over-exposure can result in liver and kidney damage.

Chronic Carbon tetrachloride can also cause damage to the kidneys and liver; in the long term it is more toxic than trichloroethylene. An early sign of kidney damage may be detected by urine examination for protein and cells. Liver damage may be indicated early by special tests or later by the appearance of jaundice. It is also under suspicion as a carcinogen.

2.7.5 Other common solvents

2.7.5.1 Benzene (C_6H_6)

MEL-TWA 5 ppm. This excellent solvent is seldom used today because of its toxic effects. It may be inhaled or absorbed via the skin and is readily absorbed by fatty tissues. A large proportion of benzene which enters the body is stored in the bone marrow which it may damage, causing anaemia or more rarely leukaemia. Benzene is altered chemically in the body and then excreted in the urine. For exposures about the MEL, blood benzene is a useful measurement. For lower exposures, breath benzene is suitable. Urinary excretion as a 'phenol' test is no longer recommended.

2.7.5.2 Toluene ($C_6H_5CH_3$) (methylbenzene) and xylene ($C_6H_4(CH_3)_2$)

Toluene, OES-TWA 100 ppm and xylene, OES-TWA 1000 ppm are frequently used solvents which have toxic effects common to other solvents. They produce narcosis and can damage the liver and kidneys. Blood or breath toluene is suitable for monitoring; for xylene, urine is tested for methyl hippuric acid.

2.7.5.3 Tetrachloroethylene ($CCl_2.CCl_2$) (perchloroethylene)

OES-TWA 50 ppm. This solvent is a narcotic and may cause liver damage. Like trichloroethylene it may break down to release phosgene when exposed to naked flames or red-hot surfaces. Monitor using blood sample taken towards the end of the working week.

2.7.5.4 Trichloroethane ($CH_3.CCl_3$) (methyl chloroform)

This solvent was once regarded as one of the safest solvents but is being withdrawn to meet requirements of an EU directive[23]. Supplies will not be available after current stocks are used up.

2.7.5.5 Carbon disulphide (CS_2)

MEL-8 hour TWA 10 ppm. Carbon disulphide is an inorganic solvent used mainly in the manufacture of viscose rayon fibres. It is absorbed through the lungs and skin and is a multi-system poison affecting the brain, peripheral nerves and the heart. Monitoring is of urinary metabolites.

2.8 Gassing

2.8.1 Gassing accidents

In the UK those gassing accidents that are reported annually occur in the following approximate order of frequency:

1 Carbon monoxide
2 Chlorine*
3 Hydrochloric acid*
4 Trichloroethylene

5 Sulphur dioxide*
6 Ammonia*
7 Hydrogen sulphide*
8 Phosgene
9 Carbon dioxide
10 Nitrous fumes
11 Phosphorus oxychloride*
12 Carbon tetrachloride

*Highly soluble gases which will irritate the eyes and upper respiratory tract with the less soluble components passing further down the tract to irritate the lung tissue.

Asphyxia caused by gassing falls into two broad categories:

Simple in which oxygen in the lungs is replaced by another gas such as carbon dioxide, nitrogen or methane.

Toxic in which there is a metabolic interference with the oxygen taken up by the body. This occurs with gases such as carbon monoxide, hydrogen sulphide and hydrogen cyanide.

2.8.2 Chlorine and hydrochloric acid (HCl)

These highly irritant gases may affect the air passages and lungs causing bronchitis and difficulties in breathing due to fluid in the lungs (pulmonary oedema).

2.8.3 Carbon monoxide (CO)

This colourless odourless gas may be found wherever incomplete combustion occurs such as in motor vehicle exhausts, furnaces, steel-works and domestic boilers.

Inhalation results in a rapid rise in CO concentration in the blood within the first hour and a much slower rise thereafter. The gas is more readily absorbed by the blood's red cells to the exclusion of oxygen and so impairs the supply of oxygen to vital organs, particularly the heart.

The effects of the gas are shown in the following table:

Exposure ppm	Probable concentration of CO in blood after 1 hr exposure (carboxy haemoglobin)	Effect
200	20%	Headache, flushed appearance, breathlessness
400	40%	Dizziness
500	50%	Collapse
600	60%	Unconsciousness

2.8.4 Hydrogen sulphide (H₂S)

Occurs in sewers, oil refineries and chemical processes. Its odour of rotten eggs can be detected at concentrations of 0.3 ppm but increasing the concentrations of exposure impairs the sense of smell. Even at low concentrations the gas irritates the eyes. Higher concentrations irritate the lungs causing pulmonary oedema (although the onset may be delayed), headache, dizziness, convulsions and unconsciousness.

2.8.5 Carbon dioxide (CO₂)

This occurs in bakeries, breweries etc. and is a result of fermentation. The gas is heavier than air. Low concentrations of CO_2 increase the rate of breathing but higher levels depress respiration causing rapid unconsciousness and even death.

2.8.6 Sulphur dioxide (SO₂)

OES-8 hour TWA 2 ppm. A colourless irritant gas with a pungent smell which causes bronchitis and pulmonary oedema.

2.8.7 Nitrous fumes (commonest form NO₂)

Pungent brown fumes which cause lung irritation after a delay of a few hours. Occurs in explosions and blasting, silo storage and diesel engine exhaust.

2.8.8 Phosgene (COCl₂)

Arises from burning chlorinated hydrocarbons, e.g. trichloroethylene. Effects similar to nitrous oxide.

2.8.9 Ammonia (NH₃)

OES-8 hour TWA 25 ppm. Ammonia has a corrosive action that will burn the skin, severely irritate or burn the cornea, cause bronchitis and pulmonary oedema.

2.9 Oxygen deficiency

Normal respiration requires:

1 An adequate concentration and partial pressure of oxygen in the inspired air.

2 A clear airway to the lungs.
3 Transfer of oxygen in the air sacs to the blood.
4 The transport of oxygen by the red cells to the tissues.

Normal oxygen requirements depend on body size, activity and fitness, and interruption of the supply can occur through failure at any of the above indicated levels. Fresh air contains approximately 21% oxygen, 79% nitrogen, 0.03% carbon dioxide. Although inspired air contains 21% oxygen, that in the air sacs has only 14% which at sea level exerts sufficient partial pressure to cross the lung–blood barrier.

At altitudes above sea level the percentage of oxygen in air is unaltered, but because the barometric pressure is less, the partial pressure of oxygen drops accordingly and makes breathing more difficult. At sea level barometric pressure equals 760 mm Hg, therefore oxygen partial pressure equals $760 \times 21/100 = 160$ mm Hg[24].

In the air sacs, however, there is vapour pressure present. It equals 47 mm Hg irrespective of altitude and diminishes the effective partial pressure which the oxygen would otherwise exert. For example, in the air sacs oxygen partial pressure at sea level equals $(760–47) \times 14/100 = 100$ mm Hg. In confined spaces the oxygen concentration can fall by several means. It can be displaced by another gas, e.g. a simple asphyxiant such as carbon dioxide. In a disused and ill-ventilated coal mine the oxygen present could be used up in oxidising the coal, resulting in a condition known as 'black damp'. Combustion requires oxygen, so that in a confined space a flame will burn up the oxygen present. Similarly, oxygen can be 'combusted' by ordinary respiration of persons working in the space. Canister type respirators should not be worn in a confined space, because of the danger of a depletion of oxygen in the atmosphere; instead full breathing apparatus should be used.

The presence of disease can also embarrass breathing, as during an attack of bronchial asthma, or in pneumoconiosis, when transfer of oxygen across the lungs is impeded. In anaemia the red cell's capacity for carrying oxygen is diminished, and in heart disease the blood may be inadequately pumped around the body. A similar effect is found with carbon monoxide poisoning, in which the normal uptake of oxygen by the red cells is prevented. Each of these mechanisms results in an inadequate oxygen supply to the tissues, a condition known as anoxia.

2.9.1 Oxygen requirement

The 'average man' of 70 kg body weight requires 0.3 litres of oxygen per minute at rest, but considerably more with activity[25].

Degree of work	Oxygen requirement (l/min)
Rest	0.3
Light	0.3–1
Moderate	1–1.5
Heavy	1.5–2
Very heavy	2–6

2.9.2 Response to oxygen deficiency

At oxygen concentration of 21–18%, the fit body tolerates exercise well. Below 18% the response depends upon the severity of work undertaken. Between 18 and 17% the body will probably not be adversely affected, unless the work undertaken is heavy, when there is likely to develop oxygen insufficiency which could lead to unconsciousness. Between 17 and 16% heavy work is not possible. Light activity will result in an increase in pulse and respiration rate in order to improve oxygen supply to the tissues.

In an environment in which the oxygen concentration is diminished it is the rate of its decline which influences body response. A sudden reduction in which the partial pressure of oxygen is inadequate for it to cross the lung–blood barrier, as might occur when the oxygen supply to an aviator at very high altitude is dramatically cut off, results in convulsions and unconsciousness within a minute and, unless promptly relieved, death. A gradual reduction in oxygen concentration may be unnoticed by the victim, there being at first a feeling of well-being and overconfidence. Then mistakes in thinking and action may occur until, at a level of 10% or lower, unconsciousness follows and, possibly, death. Should the oxygen level be restored and the individual recover, the incident might not be recalled and there could be a repetition of the mistakes as before[26]. Recovery may be complete, or there may be residual headache and weakness for some hours. The most sensitive tissues are the brain, heart and retina, which are liable to sustain damage.

2.10 Occupational cancer

Cancer is a disorder of cell growth. It begins as a rapid proliferation of cells to form the primary tumour (neoplasm) which is either benign or malignant. If benign it remains localised, but may produce effects by pressure on neighbouring tissue. A malignant tumour invades and destroys surrounding tissue and spreads via lymph and blood streams to distant body parts (metastasis) such as the lung, liver, bone or kidney (secondary tumours). The patient becomes weak, anaemic and loses weight (cachexia). Pneumonia is the commonest form of death. The incidence of cancer increases with age and is responsible for 24% of all deaths.

Cancer is caused either by the inheritance of an abnormal gene, or exposure to an environmental agent acting either directly or indirectly on the cell genes.

Of all cancers, less than 8% are occupational and due to chemical and physical agents (see *Table 2.1*). Occupational cancers tend to occur after a long latent period of some 10–40 years and at an earlier age than spontaneous cancers.

Some carcinogens act together (synergistically); an example is found in asbestos workers who smoke and are much more likely to develop cancer of the bronchus than those who do not.

Table 2.1 Table of some causes of occupational cancer in man

Agent	Body site affected	Typical occupation
Sunlight	Skin	Farmers and seamen
Asbestos	Lung, pleura, peritoneum	Demolition workers, miners
2-naphthylamine	Bladder	Dye manufacture, rubber workers
Polycyclic aromatic hydrocarbons	Skin, lung	Coal gas manufacture, workers exposed to tar
Hard wood dust	Nasal sinuses	Furniture manufacture
Leather dust	Nasal sinuses	Leather workers
Vinyl chloride monomer	Liver	PVC manufacture
Chromium fume	Lung	Chromate manufacture
Ionising radiations	Skin and bone marrow	Radiologists and radiographers

Identification of occupational cancer often depends in the first place on the observation of a cluster of cases, as occurred with cancer of the scrotum in chimney sweeps in 1755, skin cancer in arsenic workers in 1822 and cancer of the liver in PVC manufacture in 1930. Following observation of cases it is necessary to establish the potential link between cause and effect. This requires a descriptive study followed by a cohort or case control study.

Cancer may be suspected where the following are found:

1 Cluster of tumour in particular trades, i.e. chimney sweeps.
2 The chemical substance in use is listed in EH40[5] as 'may cause cancer'.
3 An Ames test proves positive.
4 Among heavy smokers in certain industries involving asbestos, chromate, nickel compounds, coke ovens, chloromethyl ether, uranium, arsenic trioxide etc.
5 The substance in use has a chemical structure that suggests carcinogenicity, e.g. aromatic amine and those substances listed in Appendix 9 of EH40[5].

Where carcinogenicity is suspected, the epidemiological tests outlined in section 2.2 should be carried out.

The classification of carcinogens is based on internationally agreed epidemiological and animal studies[27] and are:

Group 1   Carcinogenic to humans.
Group 2a   Probably carcinogenic to humans with sufficient evidence from animal studies.
Group 2b   Possibly carcinogenic to humans but absence of sufficient evidence from animal tests.
Group 3   Not classifiable as to its carcinogenicity to humans.
Group 4   No evidence of carcinogenicity in humans or animals.

Many chemical substances have been assigned the risk phrase 'R-45; May cause cancer' and these are listed in Appendix 9 of EH40[5].

Although the total number of deaths from cancer in this country is rising there is no evidence that the increase is due to the effect of industrial chemicals. The two most important factors leading to this increase appear to be the ever increasing number of lung cancer deaths due to smoking and fewer deaths from other causes such as infection thus putting more people at risk of developing cancer who otherwise would have died from other causes[28].

2.10.1 Angiosarcoma

Angiosarcoma is a rare 'cancer' of the liver, known to be associated with vinyl chloride monomer and, more rarely, with thorium dioxide. Much more commonly, angiosarcoma has occurred without a recognised association with any chemical. Vinyl chloride monomer (VCM) can be polymerised to form polyvinyl chloride (PVC) and was first discovered in Germany in the 1930s[29]. In 1966 VCM was known to cause bone disease, affecting the hands of Belgian autoclave workers employed in the manufacture of PVC. When, in 1971, the chemical was given to animals to reproduce the bone disease, it was found instead to have carcinogenic properties.

2.10.2 Vinyl chloride monomer (VCM) ($CH_2 = CH$)

This gas is polymerised when heated under pressure (i.e. molecules of the gas are joined together in long chains) to form polyvinyl chloride (PVC). Although the explosive dangers of the gas have long been recognised, it was not until 1974 that three cases in American factory workers who were making PVC from VCM indicated that it could cause a rare liver tumour, angiosarcoma. Symptoms include abdominal pain, impaired appetite, loss of weight, distention of abdomen, jaundice and death. A Code of Practice[30] gives useful guidance on the control of this substance in the work environment.

2.11 Physical agents

In recent years there has been an increasing recognition of the harm that physical agents can do to the health of people at work. Injuries from this source now account for two-thirds of the new successful claims for industrial disease compensation.

2.11.1 Hand–arm vibration syndrome (HAVS)

HAVS follows from exposure to vibrations in the range 2–1500 Hz which causes narrowing in the blood vessels of the hand, damage to the nerves

Table 2.2. Stockholm scale for the classification of the hand–arm vibration syndrome

Stage	Grade	Description
1. Vascular component		
1	Mild	Occasional blanching attacks affecting tips of one or more fingers
2	Moderate	Occasional attacks distal and middle phalanges of one or more fingers
3	Severe	Frequent attacks affecting all phalanges of most fingers
4	Very severe	As in 3 with trophic skin changes (tips)
2. Sensorineural component		
0_{SN}	–	Vibration exposed. No symptoms
1_{SN}	–	Intermittent or persistent numbness with or without tingling
2_{SN}	–	As in 1_{SN} with reduced sensory perception
3_{SN}	–	As in 2_{SN} with reduced tactile discrimination and manipulative dexterity

The staging is made separately for each hand.

and muscle fibres and to bones and joint[31] evidenced by pain and stiffness in the joints of the upper arm. The impaired circulation of blood to the fingers leads to a condition known as *vibration white finger* (VWF). The most damaging frequency range is 5–350 Hz.

2.11.1.1. Vibration white finger

There is a latent period from first exposure to the onset of blanching which can vary from one to several years depending on the magnitude and frequency of the vibration and the length of exposure. Other symptoms of numbness and tingling, which variably affect the fingers extending from the tips; coldness, pain and loss of sensation may follow. Later, there may be loss of finger dexterity (e.g. picking up objects and fastening buttons) and impairment of grip. Eventually the finger tips become ulcerated and gangrenous. The vascular and nervous effects may develop independently but usually occur concurrently. Disability is graded in accordance with the Stockholm scale (see *Table 2.2*).

2.12 Ionising radiations

Ionising radiations are so called because they produce 'ions' in irradiated body tissue. They also produce 'free radicals' which are parts of the molecule, electrically neutral but very active.

The biological consequences of radiation depend on several factors:

1 The nature of the radiation – some radiations being more damaging than others. Alpha particles are not harmful until they enter the body

by inhalation, ingestion or via a wound. Beta particles can penetrate the skin to about 1 cm and cause a burn. X-rays, gamma rays and neutrons can pass right through the body and cause damage on the way.
2 The dose and duration of exposure.
3 The sensitivity of the tissue.
4 The extent of the radiation.
5 Whether it is external or internal.

2.12.1 Sensitivity of tissue

Tissues vary in their sensitivity to radiation, the most sensitive being the lymphocytes of the blood: they respond to excess radiation by a drop in their number within a couple of days, followed by a fall in other blood cells. Next in sensitivity are the cells of the gonads, the bowel lining, the skin, lung, liver, kidney, muscle and nerves.

2.12.2 Extent of radiation

Localised radiation is generally less immediately serious than whole body radiation for the same total dose.

2.12.3 Localised external radiation effects

Exposure to a small area of the body may result in redness of the skin, or even a blister, which either heals or ulcerates. The hands are very susceptible to localised radiation, the fingers becoming swollen and tender and, if the blood vessels are affected, gangrene could develop: the nails may become ridged and brittle. Exposure to the eyes in a dose of about 2 sievert may lead to cataract after a lapse of about two years. Exposure to the gonads can cause mutation and loss of fertility.
   Injury with a threshold and dose related severity is termed *non-stochastic*; while injury with no threshold and of a random nature, as in neoplasm and DNA damage, is called *stochastic*.

2.12.4 Whole body external radiation effects

Dose		Effect
rem	Sv	
Up to 25	Up to 0.25	Probably none. Lymphocyte count might fall in two days. Sperms and chromosomes may be damaged.
25–100	0.25–1.00	Damage more likely. Drop in total white cell count.
100–200	1.00–2.00	Nausea, vomiting, diarrhoea.
200–500	2.00–5.00	Above effects plus increasing mortality.
500–1000	5.00–10.00	Rapid onset of above symptoms, shock and coma.

2.12.5 Acute radiation syndrome

A dose of some 2 sievert or more to the whole body may give rise to an 'acute radiation syndrome'. The response, depending on the intensity of the dose, begins with vomiting and diarrhoea within a few hours. By the second or third day there is an improvement, but the blood count falls. By the fifth day there is a return of symptoms, with fever and infection.

2.12.6 Internal radiation

These effects depend upon the nature of the radioactive material, its route of entry and concentration in a particular tissue, and due mainly to α or β particles. Lung cancer has been observed in miners following inhalation of radon, and severe anaemia and bone tumour following ingestion of radium in luminising dial painters.

2.12.7 Long-term effects

These may take several years to develop. Cancer of the skin or other organs has a peak incidence about seven years after exposure. The blood can be affected in two ways, either by leukaemia, which is a cancer of the white cells or, less commonly, by a severe anaemia in which the bone marrow fails to produce red cells. Chronic ulceration, loss of hair, cataracts, loss of fingertips, diminished fertility, and mutations may also occur. The maximum permitted doses are indicated in *Table 2.3*.

Table 2.3 Radiation dose limit

Body part	Dose limit per calendar year mSv
Whole body	50.0
Individual organs and tissues	500.0
Lens of eye	150.0
Woman of childbearing age	13 per 3 months
Pregnant woman	10 for period of pregnancy
Trainees under 18 years old:	
whole body	15
individual organs and tissues	150
lens of the eye	45

2.12.8 Medical examinations

A pre-employment medical is required for employees likely to receive a dose of ionising radiation exceeding three-tenths of the relevant dose limit. The examination will include a test of blood.

A certificate issued by the examining Employment Medical Adviser or factory doctor will be valid for one year.

2.12.9 Principles of control

The following simple precautions should be adopted to reduce to a minimum hazards from the use of radioactive materials:

1 Employ the smallest possible source of radiation.
2 Ensure the greatest distance between source and person.
3 Provide adequate shielding between source and person.
4 Reduce exposure time to a minimum.
5 Practise good personal hygiene where there is risk of absorption of radioactive material.
6 Personal sampling by use of (a) film badge and/or (b) thermal luminescent dose meter.
7 A dose of 15 mSv whole body in a year requires investigation of work exposure and control procedures. A cumulative dose of 75 mSv within five years requires further investigation of work, personal circumstances, dose history and advice regarding further exposure to ionising radiations.

2.13 Noise-induced hearing loss

2.13.1 Mechanism of hearing

What we perceive as sound is a series of compressions and rarefactions transmitted by some vibrating source and propagated in waves through the air[32]. The compressions and rarefactions impinge on the eardrum (tympanic membrane) causing it to vibrate and transfer the movements through three small bones in the middle ear to the fluid of the inner ear. There they are received by rows of hairs (in the organ of corti), which vary in their response to different frequencies of sound, and are then transmitted to the brain and interpreted as sound.

2.13.2 Sensitivity of the ear

The ear can interpret frequencies between 20 and 20 000 Hz approximately. Frequencies below (infrasonic) and above (ultrasonic) this range are not heard. The range of frequency for speech is between 400 and 4000 Hz.

2.13.3 Definition and effects

Noise is commonly defined as unwanted sound. The definition is dependent on individual interpretation and may or may not include the recognition that some sounds produce harmful effects. Some 'sounds' cause annoyance, fright, or stress; others may interfere with communication. Loud sounds can cause deafness. 'Noise'-induced deafness is of two kinds: temporary and permanent.

2.13.4 Temporary deafness

Exposure to noise levels of about 90 dBA for even a few minutes may induce a temporary threshold shift (change of the threshold at which sound can just be heard), lasting from seconds to hours, and which can be detected by audiometry. Temporary threshold shift (TTS) may be accompanied by 'noises' in the ears (tinnitus) and may be a warning sign of susceptibility to permanent threshold shift (PTS) which is an irreversible deafness.

2.13.5 Permanent deafness

The onset of permanent deafness may be sudden, as with very loud explosive noises, or it may be gradual. A gradual onset of deafness is more usual in industry and may be imperceptible until familiar sounds are lost, or there is difficulty in comprehending speech. The consonants of speech are the first to be missed: f, p, t, s and k. These are of high frequency compared with the vowel sounds, which are of low frequency. Speech can still be heard, but without the consonants it is unintelligible. There is a risk too that a person exposed to excessive noise may believe himself to be adjusting to it when, in fact, partial deafness has already developed.

2.13.6 Limit of noise exposure

As noise effects are cumulative, the noise emission levels should be below 85 dBA. If this is not possible they should be reduced to the lowest level possible and suitable hearing protection provided. Ten years' exposure at 90 dBA ($L_{EP.d}$) can be expected to result in a 50 dB hearing loss in 50% of the exposed population.

If the noise energy is doubled, then it is increased by 3 dBA and requires a halving of the exposure time, e.g.[32]

dBA	Hours of exposure
90	8
93	4
96	2
99	1
102	½
105	¼

The above table is helpful provided the noise level remains constant. For variable noise exposure, however, the daily personal noise exposure ($L_{EP.d}$) must be calculated.

Individuals exposed to 85 dBA must be offered hearing protection, but at 90 dBA or more hearing protection must be provided and worn.

2.13.7 The audiogram

An audiogram (*Figure 2.2*) is a measure, over a range of frequencies, of the threshold of hearing at which sound can just be detected. Early deafness occurs in the frequency range 2–6 kHz and is shown typically as a dip in the audiogram at 4 kHz. The depth of the dip depends on the degree of hearing damage and, as this worsens, so the loss of hearing widens to include neighbouring frequencies. The advantages of an audiogram are that it:

1 provides a base line for future comparison;
2 is helpful in job placement; and
3 can be used to detect early changes in hearing and in the diagnosis of noise-induced deafness.

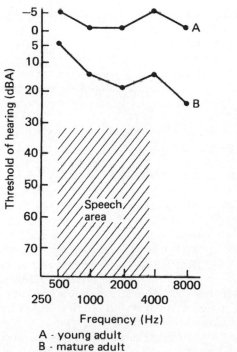
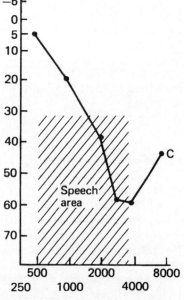

A - young adult
B - mature adult

C - adult after many years exposure to occupational noise

Figure 2.2 Audiograms

2.13.8 Occupational deafness

Disablement benefit may be awarded if deafness follows from:

1 employment in a prescribed occupation for at least 10 years, or an incident at work, and
2 the hearing loss is 50 dB averaged over the frequencies 1, 2 and 3 kHz in each ear.

2.14 Working in heat

Normally the human body maintains its core temperature within the range 36–37.4°C by balancing its heat gains and losses. Maintaining an employee's health in a hot environment requires the control of air temperature and humidity, body activities, type of clothing, exposure time and ability to sweat. To sweat freely the individual must be fit, acclimatised to the heat with sufficient water intake to ensure a urine output of about 2½ pints per day. When the air temperature reaches 35°C plus, the loss of body heat is by sweating only, but this may be difficult when humidity reaches 80% or more.

Body reactions to overheating are:

● An increase in pulse rate. The rate should fall by 10 beats/minute on cessation of exposure.
● Muscle cramp due to insufficient salt intake. Exhaustion with the individual feeling unwell and perhaps confused.
● Fainting and dizziness with pallor and sweating.
● Heat stroke is the most serious with the body temperature very high, the skin dry and flushed.
● Dehydration due to insufficient fluid intake. Prolonged dehydration may lead to the formation of stones in the kidney.

Following first aid care, the patient needs to be referred to a doctor.

2.15 Work related upper limb disorders (WRULD)[33]

WRULD covers a number of conditions variously known as telegraphist's, writer's or twister's cramp and tenosynovitis, all of which became prescribed diseases in 1948. Other common conditions include carpal tunnel syndrome, tennis and golfer's elbow.

The condition arises from frequent forceful repetitive arm movement. Early symptoms include aches and pain in the hands, wrists, forearm, elbows and shoulders with tenderness over the affected tendons and muscles. Following rest there is a quick recovery. If similar work is resumed too soon, there is likely to be a worsening to the second phase when, in addition to the symptoms, there appears a redness, swelling and marked limitation of movement. A longer period of rest is then required with possible splinting of the limb and injections of cortisone. If not treated in time the condition can become extremely disabling and may require surgical intervention. In the last few years incidents of WRULD have, numerically, exceeded that of any other group of diseases.

2.15.1 Prevention

Identify those jobs involving frequent prolonged rapid forceful movements, forceful gripping and twisting movements of the hand and arm, where the wrist is angled towards the little finger, the arm held above shoulder height or uncomfortably away from the body, and those where repetitive pushing, pulling and lifting are necessary.

Ensure hand tools are designed with good mechanical advantage and have a comfortable grip, are suitable for those who use them and that cutting edges are kept sharp.

Those involved in the work should be warned of the risks and trained in the correct use of the tools. Rest periods and work rotation should be introduced and piece work avoided.

Where the condition is suspected, complaints should be monitored and checks made of first aid records and absence certification. Susceptible persons should be examined by a doctor before further exposure.

2.16 Diseases due to micro-organisms

Micro-organisms include a variety of minute organisms such as viruses, bacteria, fungi and protozoa that can only be seen with the aid of a microscope. Micro-organisms gain entry to the body through the lungs, gut or breaks in the skin. If their virulence overcomes the body's defence, disease may result. The term pathogen covers all micro-organisms which cause disease. Diseases of animals transmitted to man are known collectively as zoonoses.

Micro-organisms account for about 10% of successful new occupational disease claims. Typical examples are:

Organism	Disease
Viruses:	hepatitis A and B, AIDS, orf
bacteria:	anthrax, legionella, leptospirosis, tuberculosis, tetanus, ornithosis, Q fever, dysentery
fungi:	Farmer's lung, ringworm, athlete's foot
protozoa:	malaria, amoebiasis
nematodes:	hookworm
BSE:	new variant C-JD

2.2.1 Hepatitis

Hepatitis is an inflammatory condition of the liver. When occurring at work it is usually caused by infections or toxic substances such as alcohol and organic solvents. Most commonly the cause is a virus of which there are three main kinds.

2.2.1.1 Infective hepatitis or 'Hepatitis A'

The virus is transmitted from infected stools to the mouth. After 2–6 weeks there occurs fever, nausea, abdominal pain and jaundice. Recovery usually occurs in 1–2 weeks and recurrence is rare.

Precautions to be taken include good personal hygiene, washing hands after the toilet and before handling food. There is a vaccine which gives long-term protection.

2.16.1.2 Serum hepatitis or 'Hepatitis B'

The virus is transmitted in infected blood or serum, especially among drug addicts who share needles; there is also a risk in renal dialysis units. The disease manifests itself after 2–6 months with symptoms similar to Hepatitis A but the effects are more prolonged and damaging.

Precautions that should be taken include the screening of donor blood for the presence of antigen and the non-reuse of needles and syringes. People at special risk are those who come into contact with blood or blood products and they should be immunised with Hepatitis B vaccine.

2.16.1.3 Hepatitis C

This is transmitted by the same route as Hepatitis B and poses an occupational hazard to a similar group of workers. There is no vaccine available.

2.16.1.4 AIDS

AIDS (acquired immune deficiency syndrome) is a breakdown in the body's immune system that can be suffered by both sexes. It can be transmitted from person to person in body fluids during sexual intercourse and infection can occur in transfusion of blood products, donated organs, by mother to child during childbirth or breast feeding and through the use of infected needles used for injections.

Within 12 weeks of infection, antibodies are found in the blood when the individual has become HIV+ve (human immunodeficiency virus positive). The virus can affect a variety of body tissues, particularly a white blood cell known as T4 which plays an important role in providing immunity. Individuals suffer in a variety of ways, from developing painless swellings of the glands in the neck and armpits to acute infections like 'flu from which recovery is usual. Other suffers experience night sweats, loss of weight, diarrhoea and fatigue. A more serious feature is an opportunistic infection whereby, following a failing immunity and fall in T4 count, the body becomes prone to a variety of bacteria, viruses and fungi including tuberculosis and pneumocystis. Diagnosis is difficult since the infecting organisms resist the usual investigative treatments and patients suffer a reduced life span.

2.16.2 Leptospirosis

Leptospirosis is caused by bacteria from the urine of infected rats (Weil's disease); dogs (*L. canicola*); cattle (*L. hardjo*) which enter via a break in the skin or into the gut if ingested. A week later a 'flu-like disease occurs sometimes with jaundice. The liver and kidneys may be severely damaged with a 15% mortality. Those most at risk include workers in abattoirs, sewers, mines, tunnels, canals, veterinary workers and those taking part in inland water sports. They should be informed of the risk and issued with a Weil's disease warning card to be presented to their doctor.

2.16.3 Legionnaire's disease

In 1976 nearly 200 American Legionnaires attending a convention at a hotel in Philadelphia collapsed with a 'flu-like disease, some with pneumonia, and there were 29 deaths. The disease was later attributed to bacteria, named *Legionella*, of which there are several types differing in pathogenicity.

Legionella bacteria occur in soils, rivers and streams, but the recent dramatic appearance of the disease in hotels, hospitals and industry is related to modern building design which allows water in air-conditioning and water systems to stagnate. At temperatures of 20–50°C bacterial growth is encouraged, which if released as a spray and inhaled, leads to pneumonia. Middle-aged smokers are most vulnerable. The disease has a 15% mortality. Diagnosis is confirmed by the presence of antibodies in the blood.

Water systems and bath shower heads should be cleaned and chlorinated periodically and a record maintained.

Pontiac fever is a less serious non-pneumonic form of the disease.

2.16.4 Anthrax

Anthrax is a highly infectious disease of ruminants: goats, cattle, sheep and horses, and is due to a bacillus. Man can be affected by direct contact with the animal, or indirectly by contact with the animal products. The disease is rare among animals in Britain, but it can be introduced into the country by infected materials, such as hides and skin, hair and wool, dried bones and bone-meal, hooves and horn. At risk particularly are those engaged in tanning, wool sorting, manufacture of brushes, bone-meal, fertiliser and glue. Also at risk are dockers and agricultural workers.

2.16.4.1 Symptoms

The disease involves the skin in about 95% of cases – the bacillus entering through an abrasion, commonly on the arm. In 2–5 days there appears a red-brown spot or pimple, which becomes a black ulcer surrounded by tiny blisters and inflamed tissue. It is usually painless, but the individual feels unwell with fever, headache, sickness and swollen glands, usually under the arm or in the groin if the leg is infected. Should the bacilli be inhaled, there follows a severe pneumonia with cough and blood-stained sputum. The mortality rate is high. Abdominal infection following ingestion is very rare. Fortunately, the disease responds to an antibiotic like penicillin if given early.

2.16.4.2 Prevention

Employees in the risk industries should be informed of the danger and carry with them an HSE card MS(B)3. All cuts must be treated and covered with a dressing while at work. Attention must be given to personal hygiene, washing hands, arms and face before meals and at end of shift. Protection can be provided by immunisation, which requires three injections at three-week intervals, a fourth six months later and then annually. Protective gloves should be worn wherever possible.

2.16.5 Humidifier fever

Humidifier fever is a 'flu-like condition which follows the inhalation of a variety of organisms such as amoeba, bacilli and fungi that grow in humidifying system. Symptoms of cough, limb pain and fever occur within a few hours of starting work. The disease is usually short term with recovery by the next day, but symptoms are likely to recur on returning to work after a few days off. It is sometimes known as Monday Morning Fever.

Water systems need to be cleaned and chlorinated periodically and a record of this maintained.

2.16.6 Tuberculosis

The incidence of tuberculosis is increasing in some communities owing to resistance to the drugs used in its treatment and to lowered resistance in AIDS patients. Infection is by inhalation of bacteria.

Mainly a disease of the lungs, its symptoms are a persistent cough, bloody sputum, night sweating and loss of weight. Sometimes no symptoms occur and the disease is first discovered on chest X-ray. Those most at risk are medical, veterinary and mortuary staff.

2.16.7 Other diseases of micro-organisms

This category has now been widened to include any infection reliably attributable to:

● work with micro-organisms
● provision of treatment for humans

- investigations involving exposure to blood and body fluids
- work with animals or any potentially infected materials so derived.

A list of the relevant conditions is given in schedule 3 of RIDDOR[34].

2.17 Psycho-social disorders

This group is probably the largest group of occupational diseases. It stems from the complex interaction of individual, social and work factors and is responsible for a great amount of sickness absence.

2.17.1 Stress

Stress is a reaction of the body to external stimuli ranging from the apparently normal to the overtly ill health. It varies with the individual personality but is one of the commonest occupational diseases.

The initial response is physiological and shows as an increase in pulse, blood pressure and respiratory rates. Although the body adjusts, persisting stimuli cause fatigue and the display of signs of 'overstress' with sweating, anxiety, tremors and dry mouth. There is difficulty in relaxing, a loss of concentration, appetite is impaired and sleep disturbed. Some may eventually become depressed, aggressive and try to avoid the cause through absenteeism or the use of alcohol or drugs. Other diseases may appear affecting the skin, peptic ulcer and coronary heart disease.

Studies have identified two personality groups: type A people who are competitive, impatient achievers and who are at greatest risk from the severe effects of stress, whereas type B people are easy-going, patient and less susceptible to pressure. Causes of stress may be considered under a number of headings:

The person   – lack of physical and mental fitness to do the job; inadequate training or skill for the particular job; poor reward and prospects; financial difficulties; fear of redundancy; lack of security in the job; home and family problems; long commuting distances.

Work demand – long hours; shift work; too fast or too slow a pace; boring repetitive work; isolation; no scope for initiative or responsibility.

Environment   – noise; heat; humidity; fumes; dust; poor ventilation; diminished oxygen; confined space; heights; poor house-keeping; bad ergonomic design.

Organisation   – poor industrial relations, welfare services and communications; inconsiderate supervision; remote management.

Common occupational causes of stress are sustained uncertainty, frustration and conflict[35]. Stress has been given a 'social rating scale'[36]

death of a spouse   100
divorce   73
marital separation   65
injury/disease   53
dismissal   47
financial difficulties   38
work responsibilities   29

2.18 Target organs

Target organs are those body parts which sustain some adverse effect when exposed to or contaminated by harmful substances or agents.

Many of the body target organs have been referred to in the text and the table below gives a summary of them with causes.

Body part	Condition	Cause
HANDS/ ARMS	Vibration white finger:	Use of vibratory tools
	Carpal tunnel syndrome:	Use of vibratory tools
	Tenosynovitis:	Repetitive pulling and twisting actions with forceful movements
	Dermatitis:	Exposure to irritants
LUNGS	Pneumoconiosis:	Mineral dust: coal, silica, asbestos, iron, tin, barium; organic dusts
	Extrinsic allergic-alveolitis:	Organic dust
	Asthma:	Proteins and low molecular weight chemicals in toxic dosages
	Irritation/ inflammation:	Nitrous fumes, phosgene, chlorine, hydrogen sulphide, sulphur dioxide, ammonia
	Infection:	Legionella, tuberculosis
	Cancer:	Asbestos, radon, nickel
SKIN	Dermatitis:	Solvents, acids/alkalis, mercury, chrome, nickel, arsenic, mineral oils, wood, plants, resin, heat
	Cancer:	Aromatic polycyclic hydrocarbons, arsenic, UV light, ionising radiations
HEAD: EARS	Deafness:	Noise
EYES	Cataracts:	Ionising radiation, UV light, heat, acids/alkalis, arc flash
	Corneal ulcers:	Ionising radiation, UV light, heat, acids/alkalis, arc flash

NOSE	Ulceration:	Chrome
TEETH	Loosening:	Mercury
	Erosion:	Sulphuric acid
	Mottling:	Fluorides
	Discoloration:	Vanadium, iodine, bromine
BRAIN	Narcosis:	Organic solvents
	Encephalop-athy:	Mercury, lead, manganese, carbon disulphide, carbon monoxide
PERIPHERAL NERVES	Neuropathy:	Lead, mercury, carbon disulphide, tetrachloroethane, trichloroethyl-ene, organo-phosphorus compounds, vibration
CARDIOVAS-CULAR	Anaemia:	Lead, arsine
	White cell count changes:	Benzene, carbon tetrachloride, ionising radiations
LIVER	Hepatitis:	Organic solvents, viruses A,B,C leptospirosis, arsenic, manganese, beryllium
	Cancer:	Hepatitis B and C, vinyl chloride monomer
KIDNEY	Toxicity:	Organic solvents, lead, mercury, cadmium
	Infection:	Micro-organisms
BLADDER	Cancer:	2-naphthylamine
BONE	Osteolysis:	Vinyl chloride monomer, vibration
	Necrosis:	Work in pressurised areas

References

1. William, P.L. and Burson, B.L. (eds) *General Principles of Toxicology, chapter 2: Industrial Toxicology,* Van Nostrand Reinhold, New York (1985)
2. Pascoe, D. *Studies in Biology, No. 149,* Edward Arnold, London (1983)
3. Harrington, J. M. and Gill, F. S., *Occupational Health Pocket Consultant,* Blackwell Scientific Publications, Oxford (1998)
4. Health and Safety Executive, *Biological monitoring in the workplace,* 2nd edn, HSE Books, Sudbury (1997)
5. Health and Safety Executive, *Guidance Note EH40, Occupational Exposure Limits,* HSE Books, Sudbury (latest issue)
6. Health and Safety Executive, Guidance Note, Environmental Health Series No. EH 64 *Occupational exposure limits: criteria document summaries.* Synopsis of the data used in setting occupational exposure limits, HSE Books, Sudbury (1994)
7. Fregert, S., *Manual of Contact Dermatitis,* A. Munksguard (1974)

8. Bishop, C. and Kipling, M. D., Dr J. Ayston Paris and Cancer of the Scrotum, 'Honour the Physician with the Honour due unto Him', *J. Soc. Occup. Med.*, **28**, 3–5 (1978)
9. Hunter, D., *The Diseases of Occupations*, 8th edn, Hodder & Stoughton, London (1994)
10. Hodgeson, G. A. and Whiteley, H. J., Distribution of pitch warts – personal susceptibility to pitch, *Brit. J. Ind. Med.*, **27**, 160–166 (1970)
11. Ref. 11, p. 20
12. Parkes, W. R., *Occupational Lung Disorders*, (3rd Edn), Butterworths, London (1990)
13. Ref. 12, p. 231
14. *Social Security (Industrial Injuries) (Prescribed Diseases) Regulations 1985*, HMSO, London (1985)
15. Alexander, W. S. and Street, A., *Metals in the Service of Man*, 6th edn, 30, Penguin Books, London (1994)
16. Waldron, H. A., Health care of people at work – workers exposed to lead, inorganic lead, *J. Soc. Occup. Med.*, **28**, 27–32 (1978)
17. Clarkson, T. W., *Mercury Poisoning Clinical Chemistry and Chemical Toxicology of Metals*, 189–204, Elsevier, Amsterdam (1977)
18. Health and Safety Executive, Mercury – medical guidance notes (Rev), Guidance Notes MS 12, HSE Books, Sudbury (1996)
19. Uvaroy, E. B., Chapman, D. R. and Isaacs, A., *Dictionary of Science*, 7th edn, Penguin Books, London (1993)
20. Matheson, D., *Occupational Health and Safety*, 2085, ILO, Geneva (1983)
21. Bird, M. G., Industrial solvents: some factors affecting their passage into and through the skin, *Annals of Occupational Hygiene*, **24**, No. 2 (1981)
22. Gompertz, D., Solvents: the relationship between biological monitoring stratagem and metabolic handling. A review, *Annals of Occupational Hygiene*, **23**, No. 4 (1980)
23. Commission of the European Communities, *Regulation No. 594/91 on banning the use of methyl chloroform*, EC Publications Department, Luxembourg and HMSO, London (1991)
24. Green, J. H., *An Introduction to Human Physiology*, 78–79, 3rd edn, Oxford University Press (1974)
25. Lamphier, E. H., *The Physiology and Medicine of Diving*, 59–60, Bailliere, Tindall & Cassell, London (1969)
26. Miles, S. and Mackay, D. E., *Underwater Medicine*, 4th edn, 107–108, Adlard Coles Ltd, St. Albans (1976)
27. International Agency for Research on Cancer, *Monograph on the evaluation of carcinogenic risks to humans*, **46**, World Health Organisation, Geneva (1989)
28. Editorial: What proportion of cancers are related to occupation?, *Lancet*, 1238, Dec. 9 (1978)
29. Gauvain, S., Vinyl chloride, *Proc. Royal Soc. Med.*, 69 (1976)
30. Health and Safety Executive, Legal guidance booklet no. L67, *Control of vinyl chloride at work (1994 edition) Control of Substances Hazardous to Health Regulations 1994 Approved Code of Practice*, HSE Books, Sudbury (1995)
31. Royal College of Physicians, *Hand Transmitted Vibrations*, 2 vols, Royal College of Physicians, London (1993)
32. Department of Employment, *Noise and the Worker*, HMSO, London (1974)
33. Health and Safety Executive, Health and Safety Guidance Booklet No. HS(G)60, *Work related upper limb disorders: a guide to prevention*, HSE Books, Sudbury (1991)
34. Health and Safety Commission, *The Reporting of Injuries, Diseases and Dangerous Occurrences Regulations 1995*, HMSO, London (1995)
35. Gross, R.D., *Psychology, the science of mind and behaviour*, 2nd edn, Hodder and Stoughton Ltd, London (1995)
36. Holmes, T.H. and Rahe, R.H., *Journal of Psychosomatic Research*, 213–218 (1967)

Further reading

National Radiological Protection Board, *Living with Radiation*, 4th edn, National Radiological Protection Board, Didcot (1989)
British Medical Association, *The BMA Guide to Living with Risk*, Penguin Books, London (1990)

Olsen, J., Merletti, F., Snashall, D. and Vuylsteek, K., *Searching for Causes of Work related Diseases*, Oxford Medical Publications, Oxford (1991)

Rose, G. and Barker, D.J.P., *Epidemiology for the Uninitiated*, British Medical Association, London (1979)

James, R.C., Industrial toxicology, Chapter 2 in *General Principles of Toxicology*, William, P.L. and Burson, B.L. eds, Van Nostrand Reinhold, New York (1985)

Pascoe, D., *Studies in Biology No. 149*, Edward Arnold, London (1983).

Chapter 3

Occupational hygiene

Dr C. Hartley

Occupational hygiene is defined by the British Occupational Hygiene Society as: 'the applied science concerned with the identification, measurement, appraisal of risk, and control to acceptable standards, of physical, chemical and biological factors arising in or from the workplace which may affect the health or well-being of those at work or in the community'.

It is thus primarily concerned with the identification of health hazards and the assessment of risks with the crucial purpose of preventing or controlling those risks to tolerable levels. This relates both to the people within workplaces and those who might be affected in the surrounding local environment.

Occupational hygiene deals not only with overt threats to health but also in a positive sense with the achievement of optimal 'comfort conditions' for workers, i.e. the reduction of discomfort factors which may cause irritation, loss of concentration, impaired work efficiency and general decreased quality of life.

The American Industrial Hygiene Association in its corresponding definition begins: 'Industrial hygiene is that science and *art* devoted to the recognition, evaluation and control . . .' [author's emphasis] indicating that although much of occupational and industrial hygiene is underpinned by proven scientific theory, a considerable amount relies on 'rule of thumb'; thus in the practical application of occupational hygiene, judgemental and other skills developed by the experienced practitioner are important.

3.1 Recognition

People at work encounter four basic classes of health hazard, examples of which are given in *Table 3.1.*

Table 3.1. Classes of health hazard

Health hazard	Example
Chemical	Exposure of worker to dusts, vapours, fumes, gases, mists etc. 100 000 chemicals are believed to be in common use in the UK at present
Physical	Noise, vibration, heat, light, ionising radiation, pressure, ultraviolet light etc.
Biological	Insects, mites, yeasts, hormones, bacteria, viruses, proteolytic enzymes
Ergonomic/ Psychosocial	Personal-task interaction, e.g. body position in relation to use of machine; harmful repetitive work. Exposure to harmful psychological stress at work

3.2 Evaluation

When a hazard in the workplace has been identified it is necessary to assess the consequent risk, interpret this against a risk tolerability standard and where appropriate apply further prevention and control measures.

3.2.1 Environmental measurement techniques

Some common environmental measurement techniques together with their interpretation as they relate to accepted standards are reviewed.

3.2.1.1 Grab sampling

'Grab sampling' is described here in the context of stain detector tubes. This involves taking a sample of air over a relatively short period of time, usually a few minutes, in order to measure the concentration of a contaminant. The results are illustrated in *Figure 3.1* where the discrete measured concentrations are plotted.

Stain detector tubes are used in this way to measure airborne concentrations of gases and vapours. Several proprietary types are available which operate on a common principle. A sealed glass tube is packed with a particular chemical which reacts with the air contaminant. The tube seal is broken, a hand pump attached, and a standard volume of contaminated air is drawn through the tube (*Figure 3.2*). The packed chemical undergoes a colour change which passes along the tube in the direction of airflow. The tube is calibrated so that the extent of colour change indicates the concentration of contaminant sampled (*Figure 3.3*).

The hand pump must be kept in good repair and recalibrated at intervals to check that it is drawing the standard volume of air and care

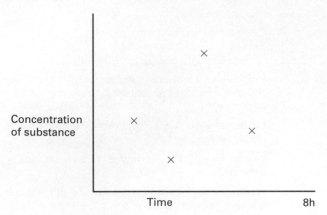

Figure 3.1 Results from grab sampling

taken to ensure that a good seal is obtained between the pump and tube.

This method of measurement has several advantages:

1 It is a quick, simple and versatile technique.
2 Stain detector tubes are available for a wide range of chemical contaminants.

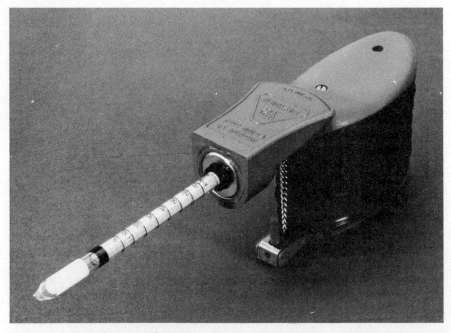

Figure 3.2 Stain detector tube and hand pump. (Courtesy Draeger Safety Ltd)

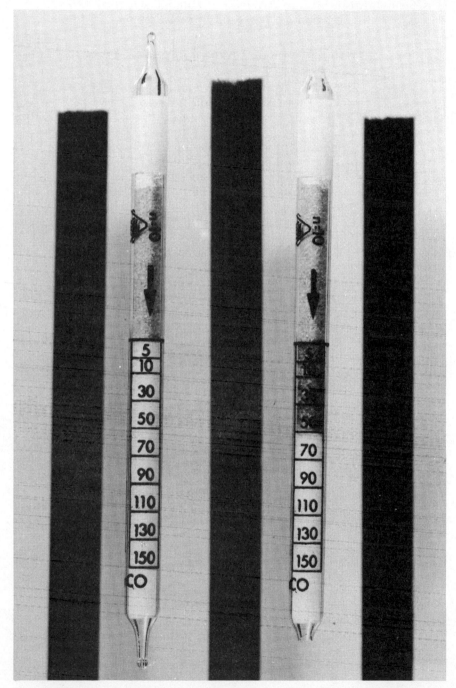

Figure 3.3 Stain detector tubes, illustrating the principle of detection. (Courtesy Draeger Safety Ltd)

3 Measurement results are provided instantaneously.
4 It is a relatively economical method of measurement.

However, it is important to be aware of the limitations of stain tubes:

1 The result obtained relates to the concentration of contaminant at the tubes inlet at the precise moment the air is drawn in. (This can be seen from Figure 3.1.) Short-term stain tubes do not measure individual worker exposure.
2 Variations in contaminant levels throughout the work period or work cycle are difficult to monitor by this technique.
3 Cross-sensitivity may be a problem since other chemicals will sometimes interfere with a stain tube reaction. For example, the presence of xylene will interfere with stain tubes calibrated for toluene. Manufacturers' handbooks give information on the known cross-sensitivities of their products. This point emphasises the need to consider the whole range of chemicals used in a process rather than just the major ones.
4 Stain tubes are not reusable.
5 Random errors associated with this technique can range up to ±25% depending on tube type.

Using a planned sampling strategy rather than an occasional tube will give a better picture of toxic contaminant levels and manufacturers will give guidance about this. However, since tubes are not reusable this method could be more costly than some of the long-term sampling techniques.

Some basic practical guidance with regard to toxic substance monitoring is given in an HSE publication[1].

There are other approaches to 'grab sampling' which involve collecting a sample of the workplace atmosphere in a suitable container, for example a special gas bag, which is taken to a laboratory for detailed chemical analysis.

3.2.1.2 Long-term sampling

This involves sampling air for several hours or even the whole work shift. The air sampling may be carried out in the worker's breathing zone (personal sampling) or at a selected point or points in the workplace (static or area sampling). Results give the average levels of contaminant across the sample period. The results are illustrated in *Figure 3.4* where the average concentration measured over the (8-hour) sampling period is X ppm.

This long-term sampling approach is discussed below for the monitoring of gas and vapour contaminants and dust and fibre aerosols. Long-term sampling methods are generally reliable, versatile and accurate, being widely used by occupational hygienists in checking compliance with hygiene standards.

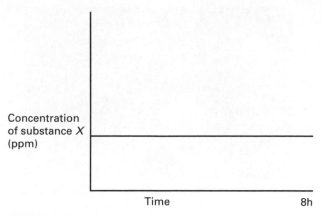

Figure 3.4 Results from long-term sampling

3.2.1.2.1 Gases and vapours

Long-term stain detector tubes are available for this purpose and are connected to a pump which draws air through the tube at a pre-determined constant rate. At the end of the sampling period the tube is examined to measure the extent of staining and hence the amount of contaminant absorbed, from which the average level of contamination can be calculated.

For some substances, direct indicating diffusion tubes are available and these devices may be used for up to 8 hours. In this case the contaminant

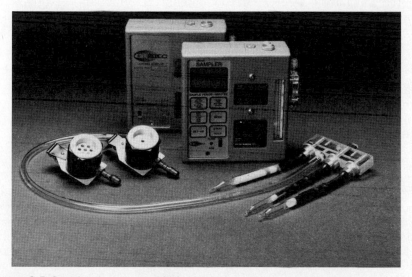

Figure 3.5 Sampling pumps suitable for both dust and vapour monitoring with collection heads. (Courtesy SKC Ltd)

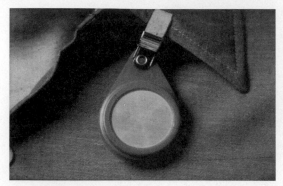

Figure 3.6 Passive monitoring badge. (Courtesy of 3M United Kingdom plc)

is collected by diffusion and no sampling pump is required – with obvious cost benefits.

Often more accurate methods are required in the assessment of worker exposure when absorbent (often charcoal) sampling is commonly used. Air is drawn by an attached low-flow portable pump through a tube containing an adsorbent (*Figure 3.5*). Further, diffusive monitors or badges containing various adsorbents are now becoming much more widely used (*Figure 3.6*). These are easy to use since they do not require a sampling pump with the result that the sampling can be carried out by suitably trained non-specialists.

Where charcoal is used, the chemical contaminant is trapped by the absorbent granules either in the tube or the diffusive monitor. After sampling is completed the tube or diffusive monitor is sealed and taken to the laboratory where the contaminant is removed (desorbed) by chemical or physical means (i.e. heat) followed by quantitative analysis, often using gas chromatography to determine the weight of contaminant absorbed during sampling.

3.2.1.2.1.1 Gas chromatography
Gas chromatography is a common and versatile analytical technique which allows both the identification and quantification of individual components of a mixture of substances. Separation of mixtures occurs after injecting a vaporised sample into a gas stream which then passes through a separating column. Components of the mixture pass through the column at different rates, depending on their individual physical properties such as boiling point and solubility, and emerge sequentially (i.e. at different times) into a detector which allows identification and quantification of each component.

This technique allows determination of the weight of contaminant adsorbed in the tube or on the diffusive monitor. When this weight is related to the known total air volume sampled (from known pump flow rate or calculated flow rate for the passive monitor and sampling time), the average airborne concentration of the substance over the sample period can be calculated.

3.2.1.2.2 Dusts and fibres

In the measurement of airborne dusts the collection device is a filter which is positioned in a holder attached to a medium flow pump. After sampling, quantification of the contaminant of the filter may be by gravimetric methods (measuring the weight change of the filter) or by using techniques such as atomic absorption spectrometry which allows the measurement of the quantity of a specific contaminant in the sample (e.g. lead). The precise analytical technique used will depend upon the part of the dust that is of interest. Also, the type of sampling device, which holds the filter, and the filter itself will depend on the nature of the dust and whether the aim is to collect the 'total inhalable' dust fraction or just the 'respirable' dust fraction.

3.2.1.2.2.1 Atomic absorption spectrometry

Atomic absorption spectrometry is a commonly used technique for the quantitative determination of metals. Absorption of electromagnetic radiation in the visible and ultraviolet region of the spectrum by atoms results in changes in their electronic structure. This is observed by passing radiation characteristics of a particular element through an atomic vapour of the sample – the sample is vaporised by aspiration into a flame or by contact with an electrically heated surface. The absorbed radiation excites electrons and the degree of absorption is a quantitative measure of the concentration of 'ground-state' atoms in the vapour. With a calibrated instrument it is possible to determine the weight of the metal captured on a filter. By knowing the total volume of air sampled, the average concentration of the metal can be calculated for the sampling period.

3.2.1.2.3 Fibre monitoring

In the measurement of airborne asbestos fibres the collection device is a membrane filter with an imprinted grid. This is positioned in an open-faced filter holder, fitted with an electrically conducting cowl and attached to a medium-flow pump. After sample collection the filter membrane is cleared, i.e. made transparent, and mounted on a micro-scope slide.

The airborne concentration of 'countable' fibres can be measured using phase contrast microscopy. 'Countable' fibres are defined as particles with length $> 5\,\mu m$, width $< 3\,\mu m$ and a length:width ratio $>3:1$. Fibres having a width of $<0.2\,\mu m$ may not be visible by the phase contrast method so this technique measures only a proportion of the total number of fibres present. However, the specified control limits take this into account.

The method requires that all fibres meeting this size definition be counted. A known proportion of the total exposed filter area is scanned using a phase contrast microscope and a tally made of *all* countable fibres in the examined area of the filter. Fibres are counted using a 'graticule' calibrated eyepiece mounted in the objective lens of the microscope. The graticule defines a known area (field) of the filter in which the fibres are counted. Knowledge of the air sample volume together with the

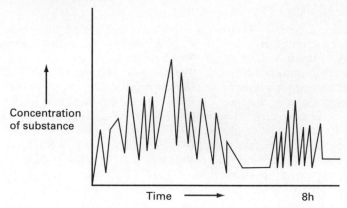

Concentration
of substance

Time ──────▶   8h

Figure 3.7 Results from direct monitoring

calculated total number of fibres on the whole filter allows calculation of the average fibre concentration over the sample period.

3.2.1.3 Direct monitoring instruments

A wide range of instruments are used in the detection of gases, vapours and dusts. These devices make a quantitative analysis giving a real-time display of contaminant level on a meter, chart recorder, data logger or other display equipment. There is a wide variety of commercially available direct monitoring instruments which are based on different physical principles of detection. Many of the physical principles involved and the range of instruments available are considered by Ashton and Gill[2]. A trace from a direct monitoring instrument is illustrated in *Figure 3.7*. From such information it is possible to work out the time weighted average concentrations during the sampling period.

Direct monitoring is particularly useful where there is a need to have immediate readings of contaminant levels, for example in the case of fast acting chemicals. It is also useful for identifying periods of peak concentration during the work cycle or work shift so that a control strategy can be developed and, for example, the point at which to install a local ventilation system can be determined. This approach is considered below.

3.2.1.3.1 Gases and vapours

Many direct reading instruments are available which are specific for particular gases, for example carbon monoxide, nitrogen oxides, hydrogen sulphide and mercury vapour. These can be linked to a chart recorder, data logger or warning device. A portable infrared gas analyser which allows direct monitoring of gases and vapours in the workplace is shown in *Figure 3.8*. In infared spectrometry the basic principle utilised is that many gases and vapours will absorb infrared radiation and under

Figure 3.8 Portable infrared gas analyser. (Courtesy Foxboro Analytical Ltd)

standard conditions the amount of absorption is directly proportional to the concentration of the chemical contaminant. This instrument takes a sample of air, detects the extent of interruption, i.e. absorption, of an infrared beam and gives a display of the contaminant concentration. Because so many substances absorb infrared radiation this is an extremely versatile instrument.

The organic vapour analyser featured in *Figure 3.9* is a portable, intrinsically safe, direct reading instrument which will monitor total hydrocarbon concentration. It will also operate as a gas chromatograph and will take a sample from the sample stream and carry out identification of mixtures and quantification of component concentrations.

3.2.1.3.2 Dusts and fibres

A 'dust lamp' is a very useful and versatile direct reading instrument. Many particles of dust are too small to be seen by the naked eye under normal lighting conditions but when a beam of strong light is passed

Figure 3.9 Organic vapour analyser. (Courtesy Quantitech Ltd)

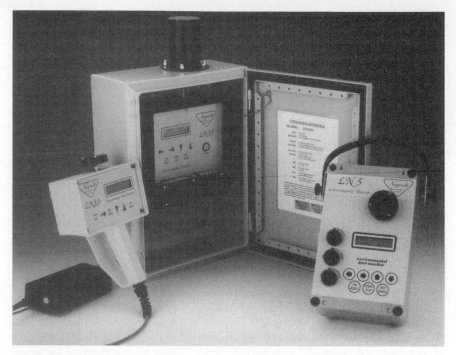

Figure 3.10 Laser nephelometer – direct reading dust monitoring instrument. (Courtesy Negretti Automation Ltd)

through a cloud of dust the particles reflect the light to the observer – known as forward light scattering – and as a result the particles are readily visible. A natural occurrence of this phenomenon is observed when a shaft of sunlight shines into a dark building highlighting the airborne particles.

Thus, if a portable lamp having a strong parallel beam is set up to shine through a dusty environment, the movement of the particles can be observed. Although this is not a quantitative method, the behaviour of the dust emitted by the work processes can be observed and corrective measures taken. It may be useful to photograph or make a video record of the observation.

Another direct reading dust monitor is the laser nephelometer (*Figure 3.10*). Ambient air is drawn continuously through a sensing chamber and illuminated by a laser. Dust particles are detected by the scattering of the laser light. This type of instrument can be used to assess PM10 and PM 2.5 concentrations in ambient air and the thoracic and respirable dust fractions in workplace atmospheres.

3.2.1.4 Oxygen analysers

Deficiency of oxygen in the atmosphere of confined spaces is often experienced in industry, for example inside large fuel storage tanks when

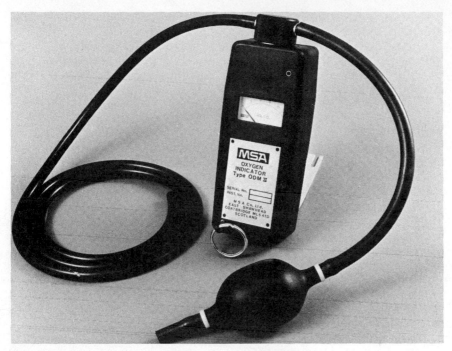

Figure 3.11 Portable oxygen analyser. (Courtesy M.S.A. Ltd)

empty. Before such places may be entered to carry out inspections or maintenance work a check must be made on the oxygen content of the atmosphere throughout the vessel. Normal air contains approximately 21% oxygen and when this is reduced to 16% or below people experience dizziness, increased heartbeat and headaches. Such atmospheres should only be entered when wearing air supplied breathing apparatus.

Portable analysers are available which measure the concentration of oxygen in the air by the depolarisation produced at a sensitive electrode mounted in the instrument (*Figure 3.11*). Several different devices are available which vary in sensitivity, reliability and ease of maintenance but they must all be checked and carefully calibrated to the manufacturers' instructions before use. Long extension probes may be attached which allow remote inspection of confined spaces.

3.2.1.5 Hygrometry

Hygrometers are instruments used in the measurement of the water vapour content of air, i.e. the humidity. The comfort and efficiency of people depends on their ability to lose heat to the environment so that body temperature may remain constant. In conditions of high temperature and humidity this heat loss cannot occur fast enough. Thus measurement of humidity and its control are important in ensuring the thermal comfort of people at work. There are also some industrial processes, e.g. textile manufacture, whose success depends on a controlled humidity in the mill.

Humidity is generally expressed as relative humidity and quoted as a percentage. It is the ratio at a given atmospheric temperature and pressure of the mass of water in a given volume of air to the mass if the air had been saturated with water.

3.2.1.6 Wet and dry bulb hygrometer

This instrument consists of two normal thermometers, one of which has its bulb exposed to the air while the other has its bulb surrounded by a wick connected to a water reservoir. Evaporation of moisture from the wick to the surrounding air causes the 'wet bulb' thermometer to show a lower reading than the corresponding 'dry bulb' thermometer. The difference between the wet and dry bulb temperature related to the dry bulb temperature defines the hygrometric state of the atmosphere. Tables have been produced, from large numbers of observations, which give the relative humidity corresponding to likely wet and dry bulb temperature combinations.

The Mason hygrometer is mounted in a static position so that readings can be taken whenever required, whilst the 'whirling hygrometer' needs to be rotated rapidly for short periods to obtain a reading.

Instruments are now available which will give a direct reading of relative humidity. In them, air is drawn across two precision matched temperature sensors; one measures dry air (dry bulb) and the other is fitted with a moistened wick (wet bulb). Signals from both sensors are processed electronically to give a direct digital readout of relative humidity.

3.2.2 Interpretation of results

3.2.2.1 Hygiene standards

When measurements of airborne contamination levels or other parameters have been made it is necessary to interpret results against a *standard*. This interpretation will form the basis for the control strategy. In considering the exposure of workers to chemicals two broad options may be presented:

1 A zero exposure policy.
2 Permit certain tolerable levels of exposure.

To achieve zero exposure to all workplace chemicals is an impossible objective in the light of present day industrial processes. However, this approach has been adopted in some countries for proven human carcinogens.

Since no-exposure as a general policy is not possible, hygiene limits have been introduced in an attempt to quantify 'safe' permissible levels of exposure. These are applied to the workplace environment and attempt to reconcile the industrial use of a wide range of materials with a level of protection of the health of exposed workers.

The setting of a hygiene limit is a two-stage process – it involves firstly the collection and evaluation of scientific data and secondly the decision-making process by a committee which also has to take into account socio-economic and political factors. The HSE claim that its OES type standards are solely based on considerations of health data and that it is only the MEL type of limit that takes account of health and socio-economic data[3].

3.2.2.2 Threshold Limit Values

The Threshold Limit Value (TLV) system is that developed and published in the USA by the American Conference of Governmental and Industrial Hygienists (ACGIH), which in spite of its title is a non-governmental organisation. It is a scientific society similar to the British Occupational Hygiene Society.

The preface to 1997 ACGIH List states:

> 'Threshold Limit Values refer to airborne concentrations of substances and represent conditions under which it is believed that nearly all workers may be repeatedly exposed day after day without adverse effect'[4]

Because of individual variation in susceptibility some workers will suffer effects ranging from discomfort to sensitisation to chemicals and occupational disease at exposure levels well below the TLV. The basis for TLVs is intended to be reasonable freedom from irritation, narcosis, nuisance or impairment of health for the majority of workers. Reference to TLVs relates to these US based standards and this system is widely used around the world.

3.2.2.3 UK exposure limits

In the past the HSE has incorporated the TLV system in a Guidance Note[5] which was published annually. However, in the mid-1980s the HSE first published a list of British exposure limits to chemical substances[3] which, in their present form, are part of the requirements of the Control of Substances Hazardous to Health Regulations 1994 (COSHH)[6,7]. These standards are used for determining adequate control of exposure (by inhalation) to hazardous substances.

When applied, the limits should not be used as an index of relative hazard and toxicity, nor should they be used as the dividing line between 'safe' and 'dangerous' concentrations. The list is not comprehensive and the absence of a substance does not indicate that it is safe.

The list of exposure limits is divided into two basic parts – namely Maximum Exposure Limits (MELs) and Occupational Exposure Standards (OESs). For substances that have been given MELs, exposure should be reduced as far as reasonably practicable, and in any case, the limit should not be exceeded. With respect to OESs, it will be sufficient to ensure that the level of exposure is reduced to the OES level. This latter requirement reflects the standard demanded for compliance with the COSHH Regulations but it should not obscure the desirable aim of reducing all exposure as far as reasonably practicable.

In order to guide the decision as to which type of standard should be assigned to an individual substance, the HSE has published its *Indicative Criteria*. With respect to OES values these are:

- the ability to identify the concentration (with reasonable certainty) at which there is no indication of injurious effects on people, with repeated daily exposure;
- the OES can reasonably be complied with; and
- reasonably foreseeable over-exposures are unlikely to produce serious short- or long-term effects on health.

An MEL value may be set for a substance which is unable to satisfy the above OES criteria and which may present serious short- and/or long-term risks to man. In some cases where a substance has been assigned an OES value, a numerically higher MEL figure may be assigned because socio-economic factors require that substance's use in certain processes.

With two exceptions, levels embodied in these values relate to personal exposure via the inhalation route, i.e. monitoring is carried out in the person's breathing zone. The list will be reprinted annually, with a list of proposed changes, together with notification of those standards which are priorities for review in the reasonably near future.

The MEL and OES occupational exposure limits are set on the recommendation of the HSC's Advisory Committee of Toxic Substances (ACTS). This follows assessment of the relevant scientific data by another committee known as the Working Group on the Assessment of Toxic Chemicals (WATCH). These committees consider both the type of limit to be set and the precise substance concentration.

The standard setting procedure is briefly outlined in the Guidance Note EH40[3]. Setting an OES is the first option and WATCH comes to a decision based upon consideration of the available information on health effects (often limited) and using the above criteria. If WATCH decides that an MEL is more appropriate, the consideration of the level at which it is to be set is dealt with by ACTS, since this involves balancing health risks against the cost of reducing exposure.

3.2.2.3.1 Maximum exposure limits (MEL)

This is the maximum concentration of an airborne substance, averaged over a reference period, to which employees may be exposed by inhalation under any circumstances. Details of the legal requirements are contained in Schedule 1 of the COSHH Regulations, as amended, and the relevant substances are listed in the first part of the HSE's Guidance Note EH40. Currently there are 55 substances listed but this is reviewed each year. A few selected examples are: hardwood dust, rubber fume, cadmium and compounds and trichloroethylene.

3.2.2.3.2 Occupational exposure standards (OES)

This is the concentration of an airborne substance, again averaged over a reference period, at which '... according to current knowledge, it is believed that there is no evidence that it is likely to be injurious to employees if they are exposed by inhalation, day after day at that concentration ...'[3]. However, current knowledge of the health effects of some chemicals is 'often limited'[3]. This is particularly the case with respect to long-term health effects on humans.

While control of exposure to the OES level satisfies minimally the requirements of COSHH Regulations, it should not discourage the application of good hygiene principles in reducing the concentration levels still further, especially in view of the limited scientific data available in respect of many of these chemicals. It would be prudent for employers to aim for concentrations of 25–50% of these levels.

3.2.2.3.3 Long-term and short-term exposure limits

Two types of exposure limit are listed with the aim of protecting against both short-term effects, such as irritation of the skin, eyes and lungs, narcosis etc., and long-term health effects. Both MEL and OES values are given as time weighted averages (TWA), i.e. the exposure concentrations measured are averaged with time over 8 hours to protect against long-term effects and over 15 minutes for protection against short-term effects.

In both the British and American systems concentrations are given in parts per million (ppm), i.e. parts of vapour or gas by volume per million parts of contaminated air, and also in milligrams of substance per cubic metre of air (mg/m^3).

3.2.2.3.4 Time weighted average concentrations (TWA)

The limits refer to the maximum exposure concentration when averaged over a 15-minute period or an 8-hour day. The time weighted average value (Cm) may be obtained from the following formula:

$$Cm = \frac{(C_1 \times t_1) + (C_2 \times t_2) + \ldots (C_n \times t_n)}{t_1 + t_2 + \ldots t_n}$$

where C_1, C_2 = concentrations measured during respective sampling periods;
t_1, t_2 = duration of sampling periods.

A simple example is where the person working an 8-hour day was exposed for 4 hours at 20 ppm vapour and then for 4 hours at 10 ppm.

$$Cm = \frac{(20 \times 4) + (10 \times 4)}{4 + 4}$$

This gives an 8-hour TWA of 15 ppm.

3.2.2.3.5 Mixtures

Most of the listed exposure limits refer to single substances or closely related groups, e.g. cadmium and compounds, isocyanates etc. A few exposure limits refer to complex mixtures or compounds, e.g. white spirit, rubber fume. However, exposure in workplaces is often to a mixture of substances and such combinations may, by their nature, increase the hazard. Mixed exposure requires assessment with regard to possible health effects, which should take into account other factors such as the primary target organs of the major contaminants and possible interaction between the latter substances.

General guidance on mixed exposures is given in EH40[3] together with a rule-of-thumb formula which may be used where there is reason to believe that the effects of the constituents of a mixture are *additive*.

$$\text{Exposure ratio} = \frac{C_1}{L_1} + \frac{C_2}{L_2} + \frac{C_3}{L_3} + \ldots$$

where C_1, C_2= time weighted average concentrations of constituents;
L_1, L_2 = corresponding exposure limits.

The use of this formula is only applicable where the additive substances have been assigned OESs. If the exposure ratio is greater than 1 then the limit for the mixture has been exceeded. If one of the substances has been assigned an MEL then the additive effect should be taken into account when deciding to what extent it is reasonably practicable further to reduce exposure.

Example
If air contains 50 ppm acetaldehyde (OES = 100 ppm) and 150 ppm secbutyl acetate (OES = 200 ppm), applying the formula:

$$\text{Additive ratio} = \frac{50}{100} + \frac{150}{200} = 1.25$$

The threshold limit is therefore exceeded.

This is a relatively crude formula and would not be applicable to a situation where two or more chemicals enhance each other's effects as is the case with synergistic reactions.

3.2.3 Physical factors

Physical factors such as heat, ultraviolet light, high humidity, abnormal pressure etc. place added environmental stress on the body and are likely to increase the toxic effect of a substance. Most standards have been set at a level to encompass moderate deviations from the normal environment. However, for gross variations, e.g. heavy manual work where respiration rate is greatly increased, continuous activity at elevated temperatures or excessive overtime, judgement must be exercised in the interpretation of permissible levels.

3.2.4 Skin absorption

Some substances have the designation 'Sk' and this refers to the potential contribution to overall exposure of absorption through the skin. In this case airborne contamination alone will not indicate total exposure to the chemical and the 'Sk' designation is intended to draw attention to the need to prevent percutaneous absorption. In the application of the assigned exposure limit it is assumed that additional exposure of the skin is prevented.

3.2.5 Sensitisation

Similarly, in the list of exposure standards[3], the designation 'Sen' is assigned to selected substances to indicate that their potential for causing sensitisation reactions has been recognised. Such substances may cause respiratory sensitisation on inhalation, for example allergic asthma, or skin effects where contact occurs, for example allergic contact dermatitis. Although not all exposed persons will become sensitised when exposed to such substances, those that do will develop ill-health effects on subsequent exposure at extremely low concentrations. Once a person has become sensitised to a substance, the occupational exposure limits are not relevant for indicating 'safe' working concentrations.

3.2.6 Carcinogens

There are differing views as to whether carcinogens should be assigned exposure limits, with one body of opinion advocating that the only safe level for a substance that can cause cancer is zero. An Approved Code of Practice[7] applies to any carcinogen which is defined in COSHH as:

- a substance classified under CHIP[8] as in the category of danger – Carcinogenic (Category 1) or Carcinogenic (Category 2);
- any substance or preparation listed in Schedule 8 of COSHH.

This definition covers substances and preparations which would require labelling with the risk phrase 'R 45 – may cause cancer' or 'R 49 – may cause cancer by inhalation.'

3.2.7 Biological agents

A major change occurred with the consolidation of COSHH in 1994 when 'biological agents' were included in the defined substances. Requirements are contained in Schedule 9 to COSHH and in a supporting Approved Code of Practice[7]. 'Biological agent' is defined in COSHH as:

> Micro-organisms, cell culture, or human endoparasite, including any which have been genetically modified, which may cause infection, allergy, toxicity or otherwise create a hazard to human health.

This refers to a general class of micro-organisms, cell cultures and human endoparasites, provided that they have one or more of the harmful properties specified in the definition. Most biologial agents are micro-organisms including bacteria, viruses, fungi and parasites. However, DNA is not to be regarded in itself as a biological agent.

Biological agents must be identified as hazards, the risks arising from work involving them assessed, and preventive or adequate control measures applied as for any other defined 'substance hazardous to health'. Biological agents are classified into four hazard groups on the following basis: their ability to cause infection, the severity of the disease that may result, the risk that such infection will spread to the community, and the availability of vaccines and effective treatment.

Key aspects are the potential of the biological agent to cause harm and the nature and degree of worker exposure to it. Risk assessments should reflect the ability they have to infect and replicate, and the possibility that there may be a significant risk to health at low exposures.

Exposure to a biological agent should be either prevented, or where this is not reasonably practicable, adequate control measures applied. Schedule 9 of COSHH is concerned with the special control provisions for biological agents. The selection of control measures should take into account the fact that biological agents do not have any air quality exposure limits and that they are able to infect and replicate at very small doses. An appropriate blend of controls is needed based on perceived levels of risk.

3.2.8 Derivation of Threshold Limit Values

Ideally, hygiene standards should be derived from the quantitative relations between the contaminant and its effects, i.e. X ppm of substances causes Y amount of harm. However, such relations are very difficult to establish in humans. The problems involved have been considered in some detail by Atherley[9]. Attempts have been made to relate human disease patterns to industrial experience, but unfortunately sufficient data do not exist. The effects of harmful agents have been studied by various methods. These include chemical analogy, which assumes that similar chemicals have biologically similar effects; and short-term testing, which may involve bacteria, animal exposure experiments and human epidemiology.

The major criteria which have been used to develop the TLV list are the effect of a substance on an organ or organ system (49%), irritation (40%), and thirdly to a lesser extent narcosis (5%) and odour (2%).

An ACGIH publication[10] summarises toxicological information on substances for which TLVs have been adopted and shows that for some substances the hazards are clear, whereas for others there is very little information on human danger. This inherent uncertainty is not reflected in the bland listing of adopted values.

In recent years, the HSE has published summaries[11] of the information used in setting its MEL and OES values. Such information may be useful in assessing the applicability of a standard to a particular workplace situation.

3.2.9 Variation in international standards

Hygiene standards vary from country to country depending upon the interpretation of scientific data and the philosophy of the regulations.

International hygiene standards for trichloroethylene illustrate this variation.

	mg/m^3	ppm
+ Australia	535	100
+ UK	535	100
+ USA (ACGIH)	267	50
+ Sweden	105	20
++ Hungary	53	10

+ Time Weighted Average.
++ Maximum Allowable Concentration.

In some countries great emphasis is put on neurophysiological changes in experimental animals as well as behavioural effects in human beings. It should be noted that although low levels may be embodied in national regulations this does not mean they are achieved in practice. In the past this has been acknowledged by former Soviet representatives. In the European Community context, there are moves to harmonise exposure limits but there are also simultaneous needs to harmonise compliance strategies.

3.2.10 Changes in hygiene limits

With new scientific evidence and changing attitudes, hygiene limits are constantly being revised. A startling example is provided by vinyl chloride monomer (VCM) which occurs in the manufacture of PVC plastics. The acute effects of VCM were identified in the 1930s as being primarily 'narcosis'. To prevent such effects during industrial use a TLV of 500 ppm was set in 1962. After further research VCM was identified as affecting the liver, bones and kidneys and the adopted value (TLV) was lowered to 200 ppm in 1971. In 1974 some American chemical workers died of a rare liver cancer (angiosarcoma) which was traced to exposure to VCM, with the result that in 1978 the adopted value (ACGIH) was dropped to 5 ppm. Hence the adopted TLV for vinyl chloride monomer has been reduced a hundredfold in under 20 years.

A section in the TLV list formally notes chemicals for which a change in the standard is intended.

Similarly, the HSE in its Guidance Note[3] publishes a list of substances where the OES is new, has been changed, or where it is intended to assign an MEL value. There is also a list of substances for which the occupational exposure limits are to be reviewed.

3.3 Control measures

When, in a workplace, a hazard has been identified and the risk to health assessed, an appropriate prevention or control strategy is then required. The general control strategy should include consideration of:

Specification.
Substitution.
Segregation.
Local exhaust ventilation.
General dilution ventilation.
Good housekeeping and personal hygiene.
Reduced time exposure.
Personal protection.

These control options should be complemented and underpinned by adequate administrative arrangements which should include the provision for regular reassessment of risks and overall review.

3.3.1 Specification

The design of a new plant or process is the ideal stage to incorporate hazard prevention and control features, e.g. limiting the quantities of toxic materials handled, the provision of remote handling facilities, utilising noise control features in the design and layout of new machinery etc. Including safety features at the design stage will be much less costly than having to add them later. This emphasises the need for safety advisers to be involved at the earliest stage of developments.

3.3.2 Substitution

This involves the substitution of materials or operations in a process by safer alternatives. A toxic material may be replaced by another less harmful substance or in another context something less flammable. An example is the widespread replacement of carbon tetrachloride by other solvents such as dichloromethane and 1,1,1-trichloroethane. In turn this substance has been phased out because of its ozone depleting characteristics and industry has found suitable substitutes.

Care needs to be exercised in the selection of 'safer' substitutes since they may be considered safer simply because there is less information available about their hazards.

Alternatively the process itself may be changed to improve working conditions with a possible benefit of increased efficiency as well.

Arc welding has been widely introduced to replace rivetting and subsequently noise levels have been reduced. Again such alterations may introduce new hazards (e.g. welding fumes) and the risks from these must be similarly assessed with the implementation of adequate controls.

3.3.3 Segregation

If a substance or process cannot be eliminated, another strategy is to enclose it completely to prevent the spread of contamination. This may be by means of a physical barrier, e.g. an acoustic booth surrounding a noisy machine or handling toxic substances in a glove box. Relocation of a process to an isolated section of the plant is another possibility that reduces the number exposed to the hazard. A particular process may be segregated in time, e.g. operated at night, when fewer people are likely to be exposed. However, in the latter case, such workers are already subjected to the additional stress of night shift working and generally function less efficiently, a point that should be borne in mind by the occupational hygienist when considering alternatives.

3.3.4 Local extract ventilation

Where it is not practicable to enclose the process totally, other steps must be taken to contain contaminants. This can be achieved by removing vapours, gases, dusts and fumes etc. by means of a local extract ventilation system. Such a system traps the contaminant close to its source and removes it so that nearby workers are not exposed to harmful concentrations.

Local extract ventilation systems have four major parts:

1 Hoods	– collection point for gathering the contaminated air into the system.
2 Ducting	– to transport the extracted air to the air purifying device or the outside atmosphere.
3 Air purifying device	– such as charcoal filters are often used to remove organic chemical contaminants.
4 Fan	– provides the means for moving air through the system.

There are several different types of local extract ventilation systems and adherence to sound design principles is necessary to achieve effective removal of contaminants. Ventilation systems must also be adequately maintained to ensure that they are operating to design specifications. This subject is dealt with in greater detail in Chapter 6.

3.3.5 Dilution ventilation

Sometimes it is not possible to extract the contaminant close to its source of origin and dilution ventilation may be used under the following circumstances where there is:

1 Small quantity of contaminant.
2 Uniform evolution.
3 Low toxicity material.

Dilution ventilation utilises natural convention through open doors, windows, roof ventilators etc. or assisted ventilation by roof fans or blowers which draw or blow in fresh air to dilute the contaminant. With both of these systems the problem of providing make-up air at the proper temperature, especially during the winter months, has to be considered.

3.3.6 Personal hygiene and good housekeeping

Both have an important role in the protection of the health of people at work. Laid down procedures are necessary for preventing the spread of contamination, for example the immediate clean-up of spillages, safe disposal of waste and the regular cleaning of work stations.

Dust exposures can often be greatly reduced by the application of water or other suitable liquid close to the source of the dust. Thorough wetting of dust on floors before sweeping will also reduce dust levels.

Adequate washing and eating facilities should be provided with instruction for workers on the hygiene measures they should take to prevent the spread of contamination. The use of lead at work is a case where this is particularly important.

Wide ranging regulations[12] and a related guidance booklet[13] dealing with workplace health, safety and welfare require that workplaces are kept 'sufficiently clean' and that waste materials are kept under control. These objectives also apply when considering other control measures.

3.3.7 Reduced time exposure

Reducing the time of exposure to an environmental stress is a control strategy which has been used. The dose of contaminant received by a person is generally related to the level of stress and the length of time the person is exposed. A noise standard for maximum exposure of people at work of 90 dBA over an 8-hour work day has been used for several years and is now contained in the Noise at Work Regulations 1989[14] as the 'second action level'. Equivalent doses of noise energy are 93 dBA for 4 hours, 96 dBA for 2 hours etc. (The dBA scale is logarithmic.) Such limiting of hours has been used as a control strategy but does not take into account the possibly harmful effect of dose rate, e.g. very high noise levels over a very short time even though followed by a long period of relatively low levels.

3.3.8 Personal protection

Making the workplace safe should be the first consideration but if it is not possible to reduce danger sufficiently by the methods outlined above the worker may need to be protected from the environment by the use of personal protective equipment. Where appropriate, the PPE Regulations require the provision of suitable PPE except where other regulations require the provision of specific protective equipment, such as the asbestos, lead and noise regulations. The PPE Regulations are supported by practical guidance[15,16] on their implementation.

Personal protective equipment may be broadly divided as follows:

1 Hearing protection.
2 Respiratory protection.
3 Eye and face protection.
4 Protective clothing.
5 Skin protection.

Personal protective devices have a serious limitation in that they do nothing to attenuate the hazard at source, so that if they fail and it is not noticed the wearer's protection is reduced and the risk he faces increases correspondingly.

Making the workplace safe is preferable to relying on personal protection; however, this regard for personal protection as a last line of defence should not obscure the need for the provision of competent people to select equipment and administer the personal protection scheme once the decision to use this control strategy has been taken. Personal protection is not an easy option and it is important that the correct protection is given for a particular hazard, e.g. ear-muffs/plugs prescribed after octave band measurements of the noise source.

Else[17] outlines three key elements of information required for a personal protection scheme:

(i)   nature of the hazard,
(ii) performance data of personal protective equipment, and
(iii) standard representing adequate control of the risk.

3.3.8.1 Nature of the danger

The hazards need to be identified and the risks assessed; for example, in the case of air contaminants the nature of the substance(s) present and the estimated exposure concentration, or, with noise, measurement of sound levels and frequency characteristics.

3.3.8.2 Performance data on personal protective equipment

Data about the ability of equipment to protect against a particular hazard is provided by manufacturers who carry out tests under controlled conditions which are often specified in national or international standards. Performance requirements for face masks, for example, are contained in two British Standards[18]. The method used to determine the noise attenuation of hearing protectors at different frequencies (octave bands) throughout the audible range is specified in a European standard[19].

3.3.8.3 Standards representing adequate control of the risk

For some risks such as exposure to potent carcinogens or protection of eyes against flying metal splinters the only acceptable level is zero. The informed use of hygiene limits, bearing in mind their limitations, would be pertinent when considering acceptable levels of air contaminants.

A competent person would need these three types of information to decide whether the personal protective equipment could *in theory* provide adequate protection against a particular hazard.

Once theoretically adequate personal protective equipment has been selected the following factors need to be considered:

1 Fit.   Good fit of equipment to the person is required to ensure maximum protection.

2 Period of use.   The maximum degree of protection will not be achieved unless the equipment is worn all the time the wearer is at risk.

3 Comfort.   Equipment that is comfortable is more likely to be worn. If possible the user should be given a choice of alternatives which are compatible with other protective equipment.

4 Maintenance.   To continue providing the optimum level of protection the equipment must be routinely checked, cleaned, and maintained.

5 Training.   Training should be given to all those who use protective equipment and to their supervisors. This should include information about what the equipment will protect against and its limitations.

6 Interference.   Some eye protectors and helmets may interfere with the peripheral visual field. Masks and breathing apparatus interfere with olfactory senses.

7 Management   This is essential to the success of personal protection commitment.   schemes.

Appropriate practice should ensure *effective* personal protection schemes are based on the requirements of regulations and codes of practice[15,16]

3.3.8.4 Hearing protection

There are two major types of hearing protectors:

1 Ear-plugs – inserted in the ear canal.
2 Ear-muffs – covering the external ear.

Disposable ear-plugs are made from glass down, plastic-coated glass down and polyurethane foam, while reusable ear-plugs are made from semi-rigid plastic or rubber. Reusable ear-plugs need to be washed frequently.

Ear-muffs consist of rigid cups to cover the ears, held in position by a sprung head band. The cups have acoustic seals of polyurethane foam or a liquid-filled annular sac.

Hearing protectors should be chosen to reduce the noise level at the wearer's ear to at least below 85 dB(A) and ideally to around 80 dB(A). With particularly high ambient noise levels this should not be done from simple A-weighted measurements of the noise level, because sound reduction will depend upon its frequency spectrum. Octave band analysis measurements[19] will provide the necessary information to be matched against the overall sound attenuation of different hearing protectors which is claimed by the manufacturers in their test data.

3.3.8.5 Respiratory protective equipment

This may be broadly divided into two types in the manner shown in *Figure 3.12*.

1 Respirators   – purify the air by drawing it through a filter which removes most of the contamination.
2 Breathing apparatus – supplies clean air from an uncontaminated source.

3.3.8.5.1 Respirators

There are five basic types of respirator:

1 Filtering Facepiece Respirator. The facepiece covers the whole of the nose and mouth and is made of filtering material which removes respirable size particles. (These should not be confused with nuisance dust masks which simply remove larger particles.)
2 Half Mask Respirator. A rubber or plastic facepiece that covers the nose and mouth and has replaceable filter cartridges.
3 Full Face Respirator. A rubber or plastic facepiece that covers the eyes, nose and mouth and has replaceable filter canisters.
4 Powered Air Purifying Respirator. Air is drawn through a filter and then blown into a half mask or full facepiece at a slight positive pressure to prevent inward leakage of contaminated air.
5 Powered Visor Respirator. The fan and filters are mounted in a helmet and the purified air is blown down behind a protective visor past the wearer's face.

Filters are available for protection against harmful dusts and fibres, and also for removing gases and vapours. It is important that respirators are never used in oxygen-deficient atmospheres.

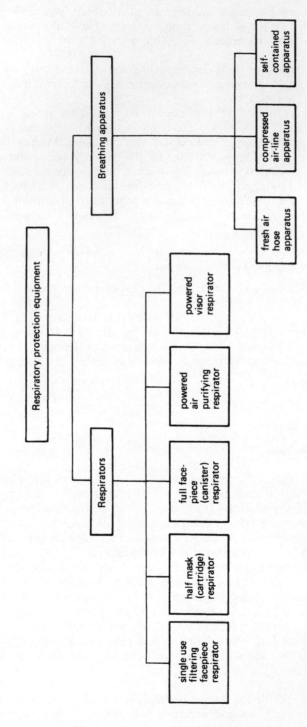

Figure 3.12 Types of respiratory protection equipment

3.3.8.5.2 Breathing apparatus

The three main types of breathing apparatus are:

1 Fresh Air Hose Apparatus. Air is brought from an uncontaminated area by the breathing action of the wearer or by a bellows or blower arrangement.
2 Compressed Air Line Apparatus. Air is brought to the wearer through a flexible hose attached to a compressed air line. Filters are mounted in the line to remove nitrogen oxides and it is advisable to use a special compressor with this equipment. The compressor airline is connected via pressure-reducing valves to half-masks, full facepieces or hoods.
3 Self-contained Breathing Apparatus. A cylinder attached to a harness and carried on the wearer's back provides air or oxygen to a special mask. This equipment is commonly used for rescue purposes.

The British Standard BS 4275[20] gives guidance on the selection, use and maintenance of respiratory protective equipment. The nominal protection factor (npf) measures the theoretical capability of respiratory protection and is used in the selection of equipment.

$$\text{npf} = \frac{\text{concentration of contaminant in the atmosphere}}{\text{concentration of contaminant in the facepiece}}$$

BS 4275 lists npf values for different devices.

3.3.8.5.3 Eye protection

After a survey of potential eye hazards the most appropriate type of eye protection should be selected. Safety spectacles may be adequate for relatively low energy projectiles, e.g. metal swarf, but for dust, goggles would be more appropriate. For people involved in gas/arc welding or using lasers, special filtering lenses would be required.

3.3.8.5.4 Protective clothing

Well-designed and properly worn, protective clothing will provide a reasonable barrier against skin irritants. A wide range of gloves, sleeves, impervious aprons, overalls etc. is currently available. The integration and compatibility of the various components of a whole-body personal protection ensemble is particularly important in high risk situations, for example in the case of handling radioactive substances or biological agents.

The factors listed above should be considered when the selection of protective clothing is being made. For example, when selecting gloves for handling solvents, a knowledge of glove material is required:

Neoprene gloves   – adequate protection against common oils, aliphatic hydrocarbons; not recommended for aromatic hydrocarbons, ketones, chlorinated hydrocarbons.

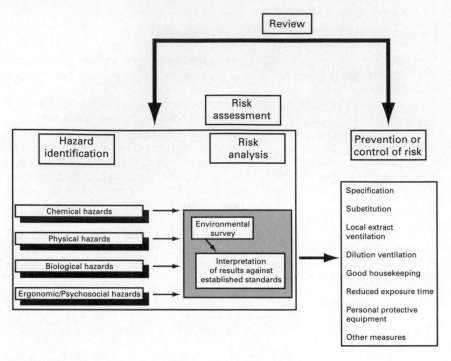

Figure 3.13 Diagram of strategy for protection against health risks

Polyvinyl alcohol gloves – protect against aromatic and chlorinated hydrocarbons.

For protective clothing to achieve its objective it needs to be regularly cleaned or laundered and replaced when damaged.

3.3.8.5.5 Skin protection

Where protective clothing is impracticable, due to the proximity of machinery or unacceptable restriction of the ability to manipulate, a barrier cream may be the preferred alternative. Skin protection preparations can be divided into the following three groups:

1 Water-miscible – protects against organic solvents, mineral oils and greases, but not metal-working oils mixed with water.
2 Water-repellent – protects against aqueous solutions, acids, alkalis, salts, oils and cooling agents that contain water.
3 Special group – cannot be assigned to a group by their composition. Formulated for specific application.

Skin protection creams should be applied before starting work and at suitable intervals during the day.

However, these preparations are only of limited usefulness as they are rapidly removed by rubbing action and care must be taken in their selection since, with some solvents, increased skin penetration can occur. The application of a moisturising cream which replenishes skin oil is beneficial after work.

3.4 Summary

The overall strategic approach is summarised in *Figure 3.13*. Although this approach to hazard identification, risk assessment and control has long been established in occupational hygiene, detailed supporting legislation has been restricted to selected health hazards, e.g. substances hazardous to health, noise, lead etc. However, new regulations[21] require this same approach to all hazards thus reinforcing hygiene practice.

References

1. Health and Safety Executive, Guidance Booklet No. HS(G) 173, *Monitoring strategies for toxic substances*, HSE Books, Sudbury (1997)
2. Ashton, I. and Gill, F.S., *Monitoring for health hazards at work*, Blackwell Scientific Publications, Oxford (1992)
3. Health and Safety Executive, Guidance Note No. EH40, *Occupational exposure limits*, HSE Books, Sudbury (This is updated annually)
4. American Conference of Governmental Industrial Hygienists, *Threshold limit values and biological indices for 1997–1998*, ACGHI, Cincinnati, Ohio (1997)
5. Health and Safety Executive, Guidance Note No. EH15/80, *Threshold limit values*, HMSO, London (1980)
6. *The Control of Substances Hazardous to Health Regulations 1994*, HMSO, London (1994)
7. Health and Safety Executive, Legal Series booklet No. L5, *General COSHH ACOP and Carcinogens ACOP and Biological Agents ACOP* (1996 edn), HSE Books, Sudbury (1997)
8. The Chemical (Hazard Information and Packaging for Supply) Regulations 1994 as amended in 1996 and 1997, HMSO, London (1997)
9. Atherley, G.R.C., *Occupational health and safety concepts*, Applied Science, London (1978)
10. American Conference of Governmental Industrial Hygienists, *Documentation of the threshold limit values and biological exposure indices*, 6th edn, ACGIH, Cincinnati, Ohio (1993)
11. Health and Safety Executive, Guidance Note No. EH64, *Summary criteria for occupational exposure limits*, 1996 with supplement in 1997, HSE Books, Sudbury (1997)
12. *Workplace (Health, Safety and Welfare) Regulations 1992*, HMSO, London (1992)
13. Health and Safety Commission, *Legislation publication No. L24 Approved Code of Practice: Workplace (Health, Safety and Welfare) Regulations 1992*, HSE Books, Sudbury (1992)
14. *The Noise at Work Regulations 1989*, HMSO, London (1989)
15. *The Personal Protective Equipment Regulations 1992*, HMSO, London (1992)
16. Health and Safety Executive, Legal Series Booklet No. L25, *Personal protective equipment at work, guidance on the Regulations* HSE Books, Sudbury (1992)
17. Else, D., *Occupational Health Practice* (ed. Schilling, R.S.F.), 2nd edn, Ch. 21, Butterworth, London (1981)
18. British Standards Institution, BS 7355 *Specification for full face masks for respiratory protective devices*, and BS 7356 *Specification for half masks and quarter masks for respiratory protective devices*, BSI, London

19. British Standards Institution, BS EN 24869–1 *Sound attenuation of hearing protectors – Methods of measurement*, BSI, London
20. British Standards Institution, BS 4275:1997 *Recommendations for the selection, use and maintenance of respiratory protective equipment*, BSI, London (1997)
21. *Management of Health and Safety at Work Regulations 1992*, The Stationery Office Ltd, London (1992)

Chapter 4

Radiation

Dr A. D. Wrixon

4.1 Introduction

Radiation is emitted by a wide variety of sources and appliances used in industry, medicine and research. It is also a natural part of the environment. The purpose of this chapter is to give the reader a broad view of the nature of radiation, its biological effects, and the precautions to be taken against it.

4.2 Structure of matter[1-3]

All matter consists of elements, for example hydrogen, oxygen, iron. The basic unit of any element is the atom, which cannot be further subdivided by chemical means. The atom itself is an arrangement of three types of particles.

1 Protons.   These have unit mass and carry a positive electrical charge.
2 Neutrons. These also have unit mass but carry no charge.
3 Electrons. These have a mass about 2000 times less than that of protons and neutrons and carry a negative charge.

Protons and neutrons make up the central part of the nucleus of the atom; their internal structure is not relevant here. The electrons take up orbits around the nucleus and, in an electrically neutral atom, the number of electrons equals the number of protons. The element itself is defined by the number of protons in the nucleus. For a given element, however, the number of neutrons can vary to form different isotopes of that element. A particular isotope of an element is referred to as a nuclide. A nuclide is identified by the name of the element and its mass, for example carbon-14. There are 90 naturally occurring elements; additional elements, such as plutonium and americium, have been created by man, for example in nuclear reactors.

If the number of electrons does not equal the number of protons, the atom has a net positive or negative charge and is said to be ionised. Thus if a neutral atom loses an electron, a positively charged ion will result. The process of losing or gaining electrons is called ionisation and occurs in the course of many chemical and physical processes.

4.3 Radioactivity[1-3]

Some nuclides are unstable and spontaneously change into other nuclides, emitting energy in the form of radiation, either particulate (e.g. α and β particles) or electromagnetic (e.g. γ-rays). This property is called radioactivity, and the nuclide showing it is said to be radioactive. Most nuclides occurring in nature are stable, but some are radioactive, for example all the isotopes of uranium and thorium. Many other radioactive nuclides (or radionuclides) have been produced artificially, such as strontium-90, caesium-137 and the isotopes of the man-made elements, plutonium and americium.

4.4 Ionising radiation[4]

The radiation emitted during radioactive decay can cause the material through which it passes to become ionised and it is therefore called ionising radiation. X-rays are another type of ionising radiation. Ionisation can result in chemical changes which can lead to alterations in living cells and eventually, perhaps, to manifest biological effects.

The ionising radiations encountered in industry are principally α, β, γ and X-rays, bremsstrahlung and neutrons. Persons can be irradiated by sources outside the body (external irradiation) or from radionuclides deposited within the body (internal irradiation). External irradiation is of interest when the radiation is sufficiently penetrating to reach the basal layer of the epidermis (i.e. the living cells of the skin). Internal irradiation arises following the intake of radioactive material by ingestion, by inhalation or by absorption through the skin or open wounds.

The α particle consists of two protons and two neutrons. It is therefore heavy and doubly charged. Alpha radiation has a very short range and is stopped by a few centimetres of air, a sheet of paper, or the outer dead layer of the skin. Outside the body, it does not, therefore, present a hazard. However, α-emitting radionuclides inside the body are of concern because α particles lose their energy to tissue in very short distances causing relatively intense local ionisation.

The β particle has mass and charge equal in magnitude to an electron. Its range in tissue is strongly dependent on its energy. A β particle with energy below about 0.07 MeV would not penetrate the outer dead layer of the skin, but one with an energy of 2.5 MeV would penetrate soft tissue to a depth of about 1.25 cm. Energy is expressed here in units of electron volt (eV), which is a measure of the energy gained by an electron in passing through a potential difference of one volt. Multiples of the electron volt are commonly used; MeV stands for Mega electron Volts (1 MeV =

1 000 000 V). As β particles are slowed down in matter, bremsstrahlung (a type of X-radiation) is produced, which will penetrate to greater distances. Thus a β-radiation source outside the body may have more penetrating radiation associated with it than is immediately apparent from the energy of the β radiation. Beta-emitting radionuclides inside the body are also of concern, but the total ionisation caused by β particles is less intense than that caused by α particles.

Gamma-rays, X-rays and bremsstrahlung are all electromagnetic radiations similar in nature to ordinary light except that they are of much higher frequencies and energies. They differ from each other in the way in which they are produced. Gamma-radiation is emitted in radioactive decay. The most widely known source of X-rays is in certain electrical equipment in which electrons are made to bombard a metal target in an evacuated tube. Bremsstrahlung is produced by the slowing down of β particles; its energy depends on the energy of the original β particles. The penetrating power of electromagnetic radiation depends on its energy and the nature of the matter through which it passes; with sufficient energy it can pass right through a human body. Sources of these radiations outside the body can therefore cause harm. With X-ray equipment, the radiation ceases when the machine is switched off. Gamma-ray sources, however, emit radiation all the time.

Neutrons are emitted during certain nuclear processes, for example nuclear fission, in which a heavy nucleus splits into two fragments. They are also produced when α particles collide with the nucleus of certain nuclides; this phenomenon is made use of in meters for measuring the moisture content of soil. Neutrons, being uncharged and therefore not affected by the electric fields around atoms, have great penetrating power, and sources of neutrons outside the body can cause harm. Neutrons produce ionisation indirectly. When a high-energy neutron strikes a nucleus in the material through which it passes, some of its energy is transferred to the nucleus which then recoils. Being electrically charged and slow moving the recoiling nucleus creates dense ionisation over a short distance.

4.5 Biological effects of ionising radiation[4-8]

Information on the biological effects of ionising radiation comes from animal experiments and from studies of groups of people exposed to relatively high levels of radiation. The best-known groups are the workers in the luminising industry early this century who used to point their brushes with the lips and so ingest radioactivity; the survivors of the atomic bombs dropped on Japan, and patients who have undergone radiotherapy. Evidence of biological effects is also available from studies of certain miners who inhaled elevated levels of the natural radioactive gas radon and its radioactive decay products.

The basic unit of tissue is the cell. Each cell has a nucleus, which may be regarded as its control centre. Deoxyribonucleic acid (DNA) is the essential component of the cell's genetic information and makes up the chromosomes which are contained in the nucleus. Although the ways in which radiation damages cells are not fully understood, many involve changes to DNA. There are two modes of action. A DNA molecule may become ionised, resulting directly in chemical change, or it may be chemically altered by reaction with agents produced as a result of the ionisation of other cell constituents. The chemical change may ultimately mean that the cell is prevented from further division and can therefore be regarded as dead.

Very high doses of radiation can kill large numbers of cells. If the whole body is exposed, death may occur within a matter of weeks: an instantaneous absorbed dose of 5 gray or more would probably be lethal (the unit gray is defined below). If a small area of the body is briefly exposed to a very high dose, death may not occur, but there may be other early effects: an instantaneous absorbed dose of 5 gray or more to the skin would probably cause erythema (reddening) in a week or so, and a similar dose to the testes or ovaries might cause sterility. If the same doses are received in a protracted fashion, there may be no early signs of injury. The effect of very high doses of radiation delivered acutely is used in radiotherapy to destroy malignant tissue. Effects of radiation that only occur above defined levels (i.e. thresholds) are known as *deterministic*. Above these thresholds, the severity of harm increases with dose.

Low doses or high doses received in a protracted fashion may lead to damage at a later stage. With reproductive cells, the harm is expressed in the irradiated person's offspring (genetic defects), and may vary from unobservable through mildly detrimental to severely disabling. So far, however, no genetic defects directly attributable to radiation exposure have been unequivocally observed in human beings. Cancer induction may result from the exposure of a number of different types of a cell. There is always a delay of some years, or even decades, between irradiation and the appearance of a cancer.

It is assumed that within the range of exposure conditions usually encountered in radiation work, the risks of cancer and hereditary damage increase in direct proportion to the radiation dose. It is also assumed that there is no exposure level that is entirely without risk. Thus, for example, the mortality risk factor for all cancers from uniform radiation of the whole body is now estimated to be 1 in 25 per sievert (see below for definition) for a working population, aged 20 to 64 years, averaged over both sexes[5]. In scientific notation, this is given as 4×10^{-2} per sievert. Effects of radiation, primarily cancer induction, for which there is probably no threshold and the risk is proportional to dose are known as *stochastic*, meaning 'of a random or statistical nature'.

4.6 Quantities and units

All new legislation in force after 1986 is required by the Units of Measurement Regulations 1980 to be in SI units. Only the SI system of units is described in full here, although the relationships between the old and new units are given in Table 4.1.

Table 4.1. Relationship between SI units and old units

Quantity	New named unit	In other and symbol	Old unit SI units	Conversion factor and symbol
Absorbed dose	gray (Gy)	Jkg^{-1}	rad (rad)	1 Gy = 100 rad
Dose equivalent	sievert (Sv)	Jkg^{-1}	rem (rem)	1 Sv = 100 rem
Activity	becquerel (Bq)	s^{-1}	curie (Ci)	1 Bq = 2.7×10^{-11}Ci

The *activity* of an amount of a radionuclide is given by the rate at which spontaneous decays occur in it. Activity is expressed in a unit called the becquerel, Bq. A Bq corresponds to one spontaneous decay per second. Multiples of the becquerel are frequently used such as the megabecquerel, MBq (a million becquerels).

The *absorbed dose* is the mean energy imparted by ionising radiation to the mass of matter in a volume element. It is expressed in a unit called the gray, Gy. A Gy corresponds to a joule per kilogram.

Biological damage does not depend solely on the absorbed dose. For example, one Gy of α radiation to tissue is more harmful than one Gy of β radiation. In radiological protection, it has been found convenient to introduce a further quantity that correlates better with the potential harm that might be caused by radiation exposure. This quantity, called the *equivalent dose*, is the absorbed dose averaged over a tissue or organ multiplied by the relevant radiation weighting factor. The radiation weighting factor for γ radiation, X-rays and β particles is set at 1. For α particles, the factor is 20. Equivalent dose is expressed in a unit called the sievert, Sv. Submultiples of the sievert are frequently used such as the millisievert, mSv (a thousandth of a sievert) and the microsievert, μSv (a millionth of a sievert).

The risks of malignancy, fatal or non-fatal, per sievert are not the same for all body tissues. The risk of hereditary damage only arises through irradiation of the reproductive organs. It is therefore appropriate to define a further quantity, derived from the equivalent dose, to indicate the combination of different doses to several tissues in a way that is likely to correlate well with the total detriment due to malignancy and hereditary damage. This quantity, derived for the fractional contribution each tissue makes to the total detriment, is called the *effective dose*. This is defined as the sum of the equivalent doses to the exposed organs and tissues weighted by the appropriate tissue weighting factor. This quantity is also expressed in sieverts.

It should be noted that the above quantities, equivalent dose and effective dose, are those defined in the latest recommendations[5] of the International Commission on Radiological Protection (ICRP). They have yet to be adopted into UK regulations, which currently use the old quantities, dose equivalent and effective dose equivalent. The differences between the old and new quantities are beyond the scope of this chapter.

4.7 Basic principles of radiological protection

Throughout the world, protection standards have, in general, been based for many years on the recommendations of the ICRP. This body was founded in 1928 and, since 1950, has been providing general guidance on the widespread use of radiation sources. The primary aim of radiological protection as expressed by ICRP[5] is to provide an appropriate standard of protection for man without unduly limiting the beneficial practices giving rise to radiation exposure. For this, ICRP has introduced a basic framework for protection that is intended to prevent those effects that occur only above relatively high levels of dose (e.g. erythema) and to ensure that all reasonable steps are taken to reduce the risks of cancer and hereditary damage. The system of radiological protection by ICRP[5] for proposed and continuing practices is based on the following general principles:

(a) No practice involving exposure to radiation should be adopted unless it produces sufficient benefit to the exposed individuals or to society to offset the radiation detriment it causes. (The justification of a practice.)
(b) In relation to any particular source within a practice, the magnitude of individual doses, the number of people exposed, and the likelihood of incurring exposure where these are not certain to be received should be kept as low as is reasonably achievable, economic and social factors being taken into account. This procedure should be constrained by restrictions on the doses to individuals (dose constraints), or risks to individuals in the case of potential exposure (risk constraints), so as to limit the inequity likely to result from the inherent economic and social judgements. (The optimisation of protection.)
(c) The exposure of individuals resulting from the combination of all the relevant practices should be subject to dose limits, or to some control of risk in the case of potential exposures. These are aimed at ensuring that no individual is exposed to radiation risks from these practices that are judged to be unacceptable in any normal circumstances. Not all sources are susceptible to control by action at the source and it is necessary to specify the sources to be included as relevant before selecting a dose limit. (Individual dose and risk limits.)

The ordering of these recommendations is deliberate; the ICRP limits are to be regarded as backstops and not as levels that can be worked up to.

For workers, the effective dose limit recommended by ICRP is 20 mSv per year averaged over defined periods of 5 years with no more than 50 mSv in any single year, the equivalent dose limit for the lens of the eye is 150 mSv in a year and that for the skin, hands and feet is 500 mSv in a year.

For comparison, the principal effective dose limit for members of the public is 1 mSv in a year. However, it is permissible to use a subsidiary dose limit of 5 mSv in a year for some years, provided that the average annual effective dose over 5 years does not exceed 1 mSv per year. The equivalent dose limits for the skin and lens of the eye are 50 mSv and 15 mSv per year respectively.

In the application of the dose limits for both workers and the public, no account should be taken of the exposures received by patients undergoing radiological examination or treatment and those received from normal levels of natural radiation.

It should be noted that these limits are in some cases more restrictive than those currently given in UK regulations (see section 4.8.1). Guidance on the implementation of the ICRP principles to the protection of workers is given in reference 9.

4.7.1 Protection against external radiation[4,6]

Protection against exposure from external radiation is achieved through the application of three principles: shielding, distance or time. In practice judicious use is made of all three. Shielding involves the placing of some material between the source and the person to absorb the radiation partially or completely. Plastics are useful materials for shielding β radiation because they produce very little bremsstrahlung. For γ and X-radiation a large mass of material is required; lead and concrete are commonly used.

Radiation from a point source falls off with the square of the distance and through absorption by the intervening air. Remote handling is one way of putting distance between the source and the person (for example, tweezers may be used when handling β-emitting sources).

4.7.2 Protection against internal radiation[4-10]

Protection against exposure from internal radiation is achieved by preventing the intake of radioactive material through ingestion, inhalation and absorption through skin and skin breaks. Eating, drinking, smoking and application of cosmetics should not be carried out in areas where unsealed radioactive materials are used. The degree of containment necessary depends on the quantity and type of material being handled: it may range from simple drip trays through fume cupboards to complete enclosures such as glove boxes. Surgical gloves, laboratory coats and overshoes may need to be worn. A high standard of cleanliness is required to prevent the spread of radioactive contamination and great care is necessary in dealing with accidental spills (*Figure 4.1*). Anyone working with unsealed radioactive material should wash and monitor his hands on leaving the working area; this is particularly important before meals are taken. Cuts and wounds should be treated immediately and no one should work with unsealed radioactive substances unless breaks in the skin are protected to prevent the entry of radioactive material.

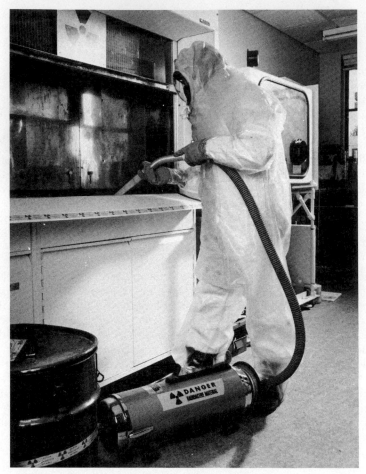

Figure 4.1 Decontamination of radioactive area in a laboratory

The radiation dose received through the intake of radioactive material depends on the mode of intake, the quantity involved, the organs in which the material becomes deposited, the rate at which it is eliminated (by radioactive decay and excretion) and the radiations emitted.

4.7.3 Radiation monitoring

The main objectives of monitoring are to evaluate occupational exposures, to demonstrate compliance with standards and regulatory requirements and to provide data needed for adequate control. For the latter, monitoring can serve the following functions:

Figure 4.2 Devices for monitoring the exposure of workers to various types of radiation. (Courtesy NRPB)

1 detection and evaluation of the principal sources of exposure,
2 evaluation of the effectiveness of radiation control measures and equipment,
3 detecting of unusual and unexpected situations involving radiation exposures,
4 evaluation of the impact of changes in operational procedures, and
5 provision of data on which the effect of future operations on radiation exposure can be predicted so that the appropriate controls can be devised beforehand and instituted.

The most appropriate means of assessing a worker's exposure to external radiations is through individual monitoring involving the wearing of a 'badge' containing radiation sensitive material, in particular a thermoluminescent chip or powder or a small piece of film (*Figure 4.2*). Doses from the intake of airborne contamination can be assessed through the use of air samplers either worn by the person or set up at appropriate points in the workplace. Radioactive material within the body can be determined by excreta or whole body monitoring, depending on the particular radionuclide involved.

The appropriate detector to be used to monitor the workplace environment depends on the type and energy of the radiation involved and whether the hazard arises from external radiation or surface or air contamination. Most survey instruments can be divided into two groups:

(a) Dose rate meters
These measure the radiation in units of dose rate and normally contain an ionisation chamber or Gieger-Müller tube. They are usually used to monitor β, γ and X-radiation fields. Special instruments are used for measuring neutron radiation dose rates.

(b) Contamination monitors
These measure the surface activity of radioactive contamination in counts per unit time. They normally contain a Geiger-Müller, proportional counter tube or scintillation counter. For α contamination, the detector normally employed would be a scintillation counter. The efficiency depends on the particular radionuclide being measured and the instrument should be calibrated for each radionuclide of interest.

The selection and use of monitoring instruments may be complex and should be discussed with a Radiation Protection Adviser (see below) or other suitable expert.

4.8 Legal requirements

The principal legislation in the UK affecting the use of ionising radiations in industry is summarised briefly below. However, readers should consult the appropriate documents for full details.

4.8.1 The Ionising Radiations Regulations 1985

These regulations, which were made under the Health and Safety at Work etc. Act 1974, came fully into effect on 1 January 1986. They apply to all work with ionising radiation rather than just work in a factory. They took account of the recommendations of ICRP that were current at the time[12] and are in conformity with a Council Directive of the European Communities which lays down basic safety standards for the health protection of the general public and workers against the dangers of ionising radiation. Details of acceptable methods of meeting the requirements of the regulations are given in the supporting Approved Code of Practice[11]. The following is a summary of some of the main requirements of the Regulations.

The dose limits for employees over the age of 18 years are those recommended by ICRP prior to 1990[12], for example, the effective dose equivalent limit for employees aged 18 years or over is 50 mSv in a year. Lower limits apply to trainees under the age of 18 years. Special restrictions apply to the rate at which women of reproductive capacity can be exposed and to the exposure of pregnant women during the declared term of pregnancy. The limits for any other person are 5 mSv in a year for the effective dose equivalent and 50 mSv in a year for the dose equivalent to individual organs or tissues other than the lens of the eye for which the value is 15 mSv in a year. The main requirement, however, is for employers to 'take all necessary steps to restrict so far as reasonably practicable the extent to which his employees and other persons are exposed to ionising radiation', in keeping with the emphasis of ICRP. If the effective dose equivalent to an employee exceeds 15 mSv in a year, the employer is required to make an investigation to determine whether it is reasonably practicable to take further steps to reduce exposure. If the effective dose equivalent exceeds 30 mSv in a calendar quarter, the dosimetry service is required to send details of the exposure to the Health and Safety Executive (HSE).

Part 4 of the Approved Code of Practice[11] requires employers to make arrangements with their dosimetry service to be informed whenever one of their employees reaches a cumulative effective dose equivalent of 75 mSv within any five consecutive calendar years. Reaching such a cumulative dose should trigger an additional investigation centred around the relevant employee. This is intended to determine what action may be necessary in the light of the risk to the individual's health from the actual and projected radiation exposure.

To facilitate the control of doses to persons, the Regulations specify criteria for designating areas as controlled or supervised areas. The underlying basis of designation is to define areas where doses may exceed three-tenths or one-tenth, respectively, of the annual dose limit for employees. Areas are designated on the basis of dose rate, total activity being handled, air activity concentration and surface contamination levels.

Employers are required to 'designate as classified workers those of his employees who are likely to receive a dose of ionising radiation which exceeds three-tenths of any relevant dose limit'. Only employees aged 18 years or over who have been certified as fit to be designated as a classified person can be so designated. Employees or other persons are only permitted to enter a controlled area if they are classified or enter under a 'written system of work'. If a written system of work is used the employer must be able to justify non-classification of the workers involved.

The Radiation Protection Adviser (RPA) is a key figure in the Regulations. An RPA is to be appointed if any employee is exposed to an instantaneous dose rate above 7.5 μSvh^{-1} or there is a controlled area which is entered. His function is to advise the employer 'as to the observance of these Regulations and as to other health and safety matters in connection with ionising radiation'. He should, for example, be consulted about restricting the exposure of workers, the identification of controlled and supervised areas, dosimetry and monitoring, the drawing up of written systems of work and local rules, the investigation of abnormally high exposures and overexposures and training. The potential appointment of an RPA should be notified to HSE one month before he takes up his role. The notification should include details of qualifications and experience and a description of the scope of the advice that the RPA will be required to give.

In relation to employees who are designated as classified persons, the Regulations require employers to ensure that assessments are made of all significant doses. For this purpose, the employer is to make suitable arrangements with an approved dosimetry service (ADS). The employer is also required to make arrangements with the ADS for that service to keep suitable summaries of any appropriate dose records for his

Figure 4.3 Checking contamination levels after a fire

employees. The purpose of the approval system is to ensure as far as possible that the doses are assessed on the basis of accepted national standards.

The Regulations also specify requirements for the medical surveillance of employees and the maintenance of individual records of medical findings and assessed doses. The general requirement to keep doses as low as reasonably practicable is strengthened by the inclusion of a basic requirement to control the source of ionising radiation and by subsequent specific requirements to provide appropriate safety devices, warning signals, handling tools etc., to leak test radioactive sources, to provide protective equipment and clothing and test them, to monitor radiation and contamination levels (see *Figure 4.3*), to store radioactive substances safely, to design, construct and maintain buildings, fittings and equipment so as to minimise contamination, and to make contingency arrangements for dealing with foreseeable but unintended incidents. In addition, undertakings holding large quantities of radioactive substances will need to make a survey of potential hazards and prepare a report, a copy of which should be sent to HSE prior to commencing work.

There are also requirements for employers to notify HSE of work with ionising radiation, overexposures and certain accidents and losses of radioactive material. The provision of information on potential hazards and appropriate training are also required. In addition, there are requirements to formulate written local rules and to provide supervision of work involving ionising radiation. This last requirement will normally necessitate the appointment by management of a radiation protection supervisor (RPS) whose responsibilities should be clearly defined.

The RPS should not be confused with the RPA. While the latter may be an outside consultant, the RPS plays a supervising role in assisting the employer to comply with the Regulations and should therefore be an employee directly involved with the work with ionising radiations, preferably in a line management position that will allow him to exercise close supervision to ensure that the work is done in accordance with the local rules, though he need not be present all the time. The RPS should therefore be conversant with the Regulations and local rules, command sufficient respect to allow him to exercise his supervisory role and understand the necessary precautions to be taken in the work that is being done.

It should be noted that the EC directive on basic safety standards referred to above has been revised in the light of the new ICRP recommendations[13]. Revision of the Ionising Radiations Regulations 1985 is currently under way; new regulations are required to be in place by May 2000.

4.8.2 The Radioactive Substances Act 1993[14]

The main purpose of this Act is to regulate the keeping and use of radioactive materials and the disposal and accumulation of radioactive waste. Under the Act those who keep or use radioactive materials on premises used for the purposes of an undertaking (trade, business, profession etc.) are required to register with the Environment Agency (England and Wales), the Scottish Environment Protection Agency or the Northern Ireland Environment and Heritage Service, according to region, unless exempt from registration. Conditions may be attached to registrations and exemptions, and these are made with regard to the amount and character of the radioactive waste likely to arise.

Furthermore, no person may dispose of or accumulate radioactive waste unless he is authorised by the appropriate Agency or Service or is exempt. Whenever possible local disposal of radioactive waste should be used but with many industrial sources, such as those used in gauges and radiography, disposal should be made through a person authorised to do so and advice should be sought from the source supplier, a Radiation Protection Adviser or the appropriate regional environment Agency or Service.

A number of generally applicable exemption orders have been made under the Act for those situations where control would not be warranted. The orders cover such things as substances of low activity, luminous articles, electronic valves, smoke detectors and some uses of uranium and thorium. The orders should be consulted for details of the conditions under which exemption is granted. The orders are currently under review.

4.8.3 Transport Regulations

Protection of both transport workers and the public is required when radioactive substances are transported outside work premises. The

Regulations and conditions governing transport in the UK and internationally follow those specified by the International Atomic Energy Agency. The latest version of the Agency's regulations is listed in reference 15. The particular regulations that apply depend on the means of transport to be used. Those that apply to the transport of radioactive materials by road are given in reference 16. These Regulations came into force on 20 June 1996 and were made under the Radioactive Material (Road Transport) Act 1991. Requirements for sending radioactive materials by post are specified in the Post Office Guide.

A full list of current regulations and guidance concerned with the transport of radioactive materials is obtainable from the Radioactive Materials Transport Division of the Department of the Environment, Transport and the Regions (tel: 0171 271 3870/3868).

4.9 National Radiological Protection Board

The National Radiological Protection Board (NRPB) was created by the Radiological Protection Act 1970. The Government's purpose in proposing the legislation was to establish a national point of authoritative reference in radiological protection.

The NRPB's principal duties are to advance the acquisition of knowledge about the protection of mankind from radiation hazards and to provide information and advice to those with responsibilities in radiological protection. Because ICRP is the primary international body to which governments look for guidance on radiation protection criteria, it is important for the UK to be in a position to influence the development of ICRP advice. A number of members of the NRPB staff are therefore actively involved in ICRP work. The NRPB also provides technical services to organisations concerned with radiation hazards, and training in radiological protection. Its headquarters are at Chilton and it has centres at Glasgow, Leeds and Chilton for the provision of advice and services. The services provided include: radiation protection adviser (RPA), reviews of design, monitoring of premises, personal monitoring, record keeping, instrument tests, testing of materials and equipment, leakage tests on sealed sources and assistance in the event of incidents and accidents. The Board runs scheduled and custom-designed training courses.

4.10 Incidents and emergencies[4,10]

In any radiological incident or emergency, the main aim must be to minimise exposures and the spread of contamination. Pre-planning against possible incidents is essential and suitable first aid facilities should be provided. Where significant quantities of radioactive substances are to be kept, procedures for dealing with fires should be discussed in advance with the local fire service.

Spills should be dealt with immediately and appropriate monitoring of the person and of surfaces should be carried out. Anyone who cuts or wounds himself when working with unsealed radioactive material must obtain first aid treatment and medical advice. This is particularly important as contamination can be readily taken into the bloodstream through cuts. If a radioactive source is lost immediate steps must be taken to locate it and, if it is not accounted for, the appropriate regional environment Agency or Service and the HSE must be notified.

The National Arrangements for Incidents involving Radioactivity (NAIR) enables police to obtain expert advice on dealing with incidents (for example, transport accidents) that may involve radiation exposure of the public and for which no other pre-arranged contingency plans exist or, for some reason, those plans have failed to function. A source of radiological advice and assistance exists in each police administrative area – hospital physicists and health physicists from the nuclear industry, government and similar establishments. The scheme is co-ordinated by the National Radiological Protection Board at Chilton from whom further details are obtainable.

4.11 Non-ionising radiation

There are several forms of non-ionising electromagnetic radiation that may be encountered in industry[17,18]. They differ from γ and X-rays in that they are of longer wavelength (lower energy) and do not cause ionisation in matter. They are ultraviolet (a few tens of nanometres (nm) to 400 nm wavelength), visible (400 to 700 nm) and infrared (700 nm to 1 mm) radiations in the optical region, and microwave and radiofrequency radiations and electric and magnetic fields. The ability of radiation within one of these defined regions to produce injury may depend strongly on the wavelength. *Figure 4.4* illustrates the monitoring for non-ionising radiation around a PVC welding machine.

4.11.1 Optical radiation

Ultraviolet radiation is used for a wide variety of purposes such as killing bacteria, creating fluorescence effects and curing inks[19]. It is produced in arc welding or plasma torch operations and is emitted by the sun. Short wavelength ultraviolet radiation of wavelength approximately less than 240 nm is strongly absorbed by oxygen in the air to produce ozone which is a chemical hazard. The OES for ozone is 0.1 ppm. Even below this level it may cause smarting of the eyes and discomfort in the nose and throat. It has a characteristic smell.

Ultraviolet radiation does not penetrate beyond the skin and is substantially absorbed in the cornea and lens of the eye. The human organs at risk are therefore the skin and the eyes. The immediate effects are erythema (as in sunburn) and photokeratitis (arc eye, snow blindness). Long-term effects are premature skin ageing and skin cancer, and possibly cataracts. No cases of skin cancer due to occupational

Figure 4.4 Characterisation of electric fields from a radio frequency PVC welding machine

exposure to artificial sources of ultraviolet radiation have been identified, but a casual link between skin cancer and exposure to solar ultraviolet radiation is now accepted, particularly for those with white skin[20]. Some chemicals such as coal tar can considerably enhance the ability of ultraviolet radiation to produce damage.

Wherever possible, ultraviolet radiation should be contained[19-22]. If visual observation of any process is required, this should be through special observation ports transparent to light but adequately opaque to ultraviolet radiation. Where the removal of covers could result in accidental injurious exposures, interlocks should be fitted which either cut the power supply or shutter the source. Protection is also achieved by increasing the distance between source and person, covering the skin and protecting the eyes with goggles, spectacles or face shields.

Intense sources of visible light such as arc lamps and electric welding units and, of course, the sun can cause thermal and photochemical damage to the eye; they can also produce burns in the skin. Adequate protection is normally achieved by keeping exposures below discomfort levels.

Infrared radiation is emitted when matter is heated. The principal biological effects of exposure can be felt immediately as heating of the skin and the cornea. Long-term exposure can cause cataracts. Protection is achieved by shielding the source and through the use of personal protective equipment especially eye wear.

The intensity of laser sources in the ultraviolet, visible and infrared regions can be orders of magnitude higher than that of other optical sources. Because of their very low beam divergence some lasers are capable of delivering large amounts of power to a distant target. Of particular importance is the injury that can be caused to the eye, such as retinal burns and cornea damage. Protection is achieved through appropriate design of equipment and through the use of a combination of the following: enclosure of the device, safety interlocks, shutters, warning signs, eye and skin protection and adequate operator training. It is also necessary to guard against stray reflections.

Lasers are widely used in the workplace for a variety of purposes ranging from cutting and welding to materials analysis and measurement. The types of laser used including their output powers vary depending on the application. The current British Standard for laster safety[21] provides appropriate advice to both the manufacturer and the user of laser products.

4.11.2 Electric and magnetic fields

Time varying electric and magnetic fields are emitted by numerous devices in the transmission and distribution of electricity and in the use of electrical and electronic equipment. Sources of exposure include power lines, induction heaters, broadcast and telecommunication systems, microwave ovens and radar[23,24].

The restriction on extremely low frequency (ELF) fields such as those generated at power distribution frequencies are based on the avoidance of induced electric currents affecting central nervous system function. At radiofrequencies which include microwaves, the fields can penetrate the body and cause heating; restrictions on exposure are designed to prevent adverse responses to increased heat load and elevated body temperatures.

In addition to the direct effects of the interaction of the electric and magnetic fields with people there is also the possibility of people touching metallic objects in the field. The indirect effect of such contact may be shock or burn. While there is no good evidence of harm to people from exposure to electromagnetic fields at existing environmental levels[25], NRPB has developed advice that is coherent in respect of the way the guideline exposure levels vary with frequency for both electric and magnetic fields[26]. The advice takes into account both direct and indirect effects.

The NRPB guidelines are based on the study of human populations. The NRPB's view is that epidemiological studies are not sufficient to form the basis for restricting human exposure to electromagnetic fields. Nevertheless, the guidelines are supported by the findings of two reports that deal with general health[25] and the risk of cancer[24].

At frequencies between 10 Hz and 1 kHz, a cautious estimate of the threshold current density to adversely affect the central nervous system is $10\,\text{mA}\,\text{m}^{-2}$. Progressively larger current densities are necessary at frequencies above and below this range.

At frequencies above 100 kHz, the restrictions on heat load are averaged over the whole body or small masses of tissue to avoid the adverse effect of localised heating. The NRPB has produced guidance based on a restriction of $0.4 \, W \, kg^{-1}$ averaged over the whole body mass. This value is considered to incorporate a sufficient margin such that it should not be necessary to account for environmental factors and work loads for healthy people.

Localised exposures are restricted to $10 \, W \, kg^{-1}$ in the head and trunk and $20 \, W \, kg^{-1}$ in the limbs, the averaging masses being 10 g and 100 g according to target tissue. Further details of the basic restrictions on exposure and derived reference levels of external electric and magnetic field strength are given in the NRPB document[25].

References

Atomic structure and radioactivity
1. Evans, R. D., *The Atomic Nucleus*, McGraw-Hill, New York (1955)
2. Royal Commission on Environmental Pollution, 6th Report, *Nuclear Power and the Environment*, Cmnd. 6618, HMSO London (1976)
3. Burchman, W. E., *Elements of Nuclear Physics*, Longman, London (1979)

Ionising radiation
4. Bennellick, E. J., Ionising radiation. In *Industrial Safety Handbook*, (Ed. W. Handley), 2nd edn, McGraw-Hill, London (1977)
5. ICRP, 1990 Recommendations of the International Commission on Radiological Protection, ICRP Publication 60, Pergamon Press, Oxford. Ann. ICRP, **21**, No. 1–3 (1991)
6. Hall, G. J., *Radiation and Life*, Pergamon Press, Oxford (1984)
7. Cox, R., Muirhead, C.R., Stather, J.W., et al., *Risk of Radiation Induced Cancer at Low Doses and Low Dose Rates for Radiation Protection Purposes*, Documents of the NRPB, Vol. 7, No. 6, NRPB, Chilton (1995)
8. Edwards, A.A. and Lloyd, D.C., *Risk from Deterministic Effects of Ionising Radiations*, Documents of the NRPB, Vol. 7, No. 3, NRPB, Chilton (1996)
9. ICRP, *General Principles for the Radiation Protection of Workers*, ICRP Publication 75, Pergamon Press, Oxford. Ann. ICRP, **27**, No. 1 (1997)
10. Martin, A. and Harbison, S.A., *An Introduction to Radiation Protection*, 4th edn, Chapman and Hall, London (1996)
11. Health and Safety Executive, Publication No. L 58, *The protection of persons against ionising radiation arising from any work activity*, HSE Books, Sudbury (1994)
12. ICRP, Recommendations of the International Commission on Radiological Protection (adopted 17 January 1977), ICRP Publication 26, Pergamon Press, Oxford. Ann. ICRP, **1**, No. 3 (1977)
13. Council of the European Communities, Council Directive 96/29/Euratom of 13 May 1996 laying down basic safety standards for the protection of the health of workers and the general public against the dangers arising from ionising radiation, *Official Journal of the European Communities*, Vol. 39, L 159, 29 June 1996
14. Department of the Environment, *Radioactive Substances Act 1993*, HMSO, London (1993)
15. International Atomic Energy Authority, IAEA Safety Standards Series publication ST-1, *Regulations for the Safe Transport of Radioactive Material*, 1996 Edn, IAEA, Vienna (1996)
16. Department of Transport, *The Radioactive Material (Road Transport) (Great Britain) Regulations 1996*, HMSO, London (1996)

Non-ionising radiation (general)
17. McHenry, C.R., Evaluation of exposure to non-ionising radiation. In Patty's *Industrial Hygiene and Toxicology, Vol. III, Theory and Rationale of Industrial Hygiene Practice* (Eds L. V. Cralley and L. J. Cralley), John Wiley & Sons, New York (1979)

18. Sliney, D.H., Non-ionising radiation. In *Industrial Environmental Health* (Eds L. V. Cralley et al.), Academic Press, London (1972)

Optical radiation
19. McKinlay, A.F., Harlen, F. and Whillock, M.J., *Hazards in Optical Radiations. A Guide to Sources, Uses and Safety*, Adam Hilger, Bristol (1988)
20. NRPB, *Health Effects from Ultraviolet Radiation*, Documents of the NRPB, Vol. 6, No. 2, NRPB, Chilton (1995)
21. British Standards Institution, *Safety of Laser Products, Part 1, Equipment Classification, Requirements and User's Guide*, BS EN 60825:1994, BSI, London (1994)
22. Health and Safety Executive, *Guidance Notes, Medical Series No. MS 15, Welding*, HSE Books, Subdury, (1978)

Electric and magnetic fields
23. Allen, S.G., Blackwell, R.P., et al., NRPB Report No. R265, *Review of Occupational Exposure to Optical Radiation and Electric and Magnetic Fields with regard to the Proposed EU Physical Agents Directive.* NRPB, Chilton (1994)
24. NRPB, *Electromagnetic Fields and the Risk of Cancer, Report of an Advisory Group on Non-ionisin Radiation*, Document of the NRPB, Vol. 3, No. 1, NRPB, Chilton (1992)
25. Dennis, A.J., Muirhead, C.R. and Ennis, J.R., NRPB Report R 241, *Human Health and Exposure to Electromagnetic Radiation*, NRPB, Chilton (1992)
26. NRPB, *Restrictions on Human Exposure to Static and Time Varying Electromagnetic Fields and Radiation*, Document of the NRPB, Vol. 4, No. 5, NRPB, Chilton (1993)

Chapter 5

Noise and vibration

R. W. Smith

The first four sections of this chapter explain what noise is, how it is defined and the theory and practice behind the measurement of noise levels. The rest outlines the way the ear works and the damage that can occur to cause noise-induced hearing loss. Some of the problems created by vibrations are considered. Reference is made to the guidelines, recommendations and legislation that exist and which are aimed at limiting the harmful effects of noise in the workplace, and the nuisance effect on the community.

5.1 What is sound?

A vibrating plate will cause corresponding vibrations or pressure fluctuations in the surrounding air, which would then be transmitted through to the receiver. For example, when an alternating electrical signal is fed into a loudspeaker, the cone vibrates causing the air in contact with it to

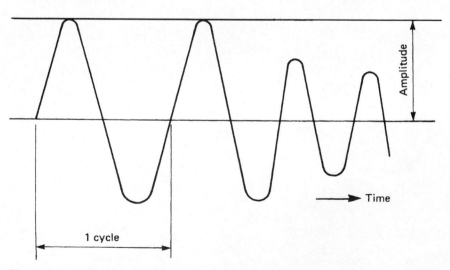

Figure 5.1 Amplitude and frequency

vibrate in sympathy, and a sound wave is produced. These pressure waves are transmitted through the air at a finite speed. This is easily demonstrated by observing the time interval between a flash of lightning and hearing a clap of thunder. The velocity of sound in air at normal temperature and pressure is approximately 342 metres per second (1122 feet per minute), at 20°C. Increasing the temperature of air increases the velocity.

These pressure fluctuations or vibrations have two characteristics: firstly the amplitude of the vibration, and secondly the frequency – both are illustrated in *Figure 5.1*.

5.1.1 Amplitude

The amplitude of a sound wave determines loudness, although the two are not directly related, as will be explained later. Typically these pressure amplitudes are very small. For the average human being the audible range is from the threshold of hearing at 20 μPa up to 200 pascals (Pa) where the pressure becomes painful. This is a ratio of 1 to 10^6. The intensity of noise is proportional to the pressure squared hence the range of intensity covers a ratio of 1 to 10^{12}.

With such a range it becomes more convenient to express the intensity of pressure amplitude on a logarithmic base. The intensity level is proportional to the square of the pressure, thus the sound pressure level (Lp) can be defined as:

$$Lp = 10 \log_{10} (P_1/P_0)^2 \tag{1}$$

where the sound pressure level (SPL) is expressed in decibels (dB), P_1 equals the pressure amplitude of the sound and P_0 is the reference pressure 20 μPa. All logarithmic calculations are to the base 10. Typical examples of sound pressure levels for a variety of environments are shown in *Figure 5.2*.

Note that, since the decibel is based on a logarithmic scale, two noise levels cannot be added arithmetically. Hence, the resultant L_{pr} from adding sources L_{p1}, L_{p2} etc. is obtained thus:

$$L_{Pr} = 10 \log_{10} \left[\left(\frac{P_1}{P_0} \right)^2 + \left(\frac{P_2}{P_0} \right)^2 + \ldots \right] \tag{2}$$

For two equal sources $L_{p1} = L_{p2}$

$$\therefore L_{pr} = 10 \log_{10} \left[\left(\frac{P_1}{P_0} \right)^2 \times 2 \right] \tag{3}$$

$$= 10 \log_{10} \left(\frac{P_1}{P_0} \right)^2 + 10 \log_{10} 2 \tag{4}$$

$$= L_p + 3.1 \text{ (dB)} \tag{5}$$

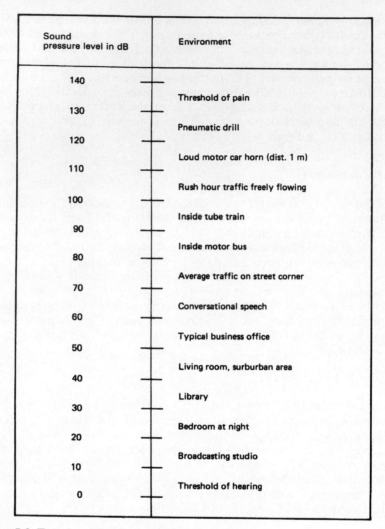

Sound pressure level in dB	Environment
140	
	Threshold of pain
130	
	Pneumatic drill
120	
	Loud motor car horn (dist. 1 m)
110	
	Rush hour traffic freely flowing
100	
	Inside tube train
90	
	Inside motor bus
80	
	Average traffic on street corner
70	
	Conversational speech
60	
	Typical business office
50	
	Living room, surburban area
40	
	Library
30	
	Bedroom at night
20	
	Broadcasting studio
10	
	Threshold of hearing
0	

Figure 5.2 Typical sound pressure levels

Thus for all practical purposes doubling the sound intensity increases the sound pressure level by 3 dB. For example:

$$90\,dB + 90\,dB = 93\,dB \tag{6}$$

Similarly, 103 dB + 90 dB = 103.2 dB, showing that where the effect of a small source is to be added to a large source, the resultant noise level would be relatively unchanged.

Of course, the converse is true, and this is important when considering the control of noise from a number of different sources, since treatment of a minor source may not result in any change in overall levels.

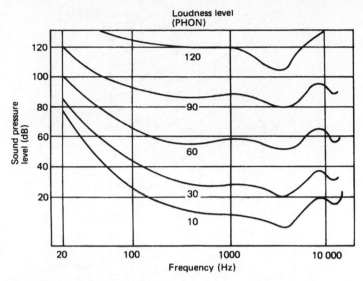

Figure 5.3 Equal loudness contours for pure tones

5.1.2 Frequency

The rate at which these pressure fluctuations take place is called frequency and is measured in hertz (Hz) or cycles/s. The human ear is normally capable of hearing over a range from 20 Hz to 16 000 Hz (16 kHz) although this range can be considerably reduced at the high frequency end for older people and for those suffering from hearing impairment.

5.2 Other terms commonly found in acoustics

5.2.1 Loudness

The human ear does not respond equally to all frequencies. To obtain the same subjective loudness at low frequencies as at higher frequencies requires a larger physical amplitude (greater L_p), since the ear is less sensitive at low frequencies. These curves of equal loudness are illustrated in *Figure 5.3*. The curves of equal loudness are defined in dB phon, and are obtained experimentally using pure tones to create the same sensation of loudness at different frequencies. From these curves it can be seen that the difference between physical amplitudes at different frequencies required to produce the same loudness curve reduces as the loudness increases.

As an approximation, an increase in sound pressure level of 8–10 dB corresponds to a subjective doubling of loudness. There are a number of different ways of calculating the loudness, and the loudness level in phon should be referenced to the method used.

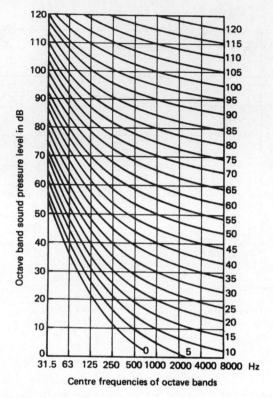

Figure 5.4 Noise rating curves

It is apparent that the subjective response to noise is extremely complex and these complexities should always be borne in mind when dealing with individual people. A noise or noise level which is acceptable to one individual may not be to another.

5.2.2 Noise rating curves

There are a standard series of Noise Rating (NR) Curves (*Figure 5.4* which are stylised forms of the loudness response curves. These NR curves are often used as a criterion for noise control, and as such are internationally accepted. Other criteria may also be encountered such as NC curves (Noise Criteria).

5.2.3 Octave bands

The previous two sections have not attempted to define frequency in terms of a bandwidth. The generally accepted bandwidth used within the

Table 5.1. Octave bands

Octave band centre frequency (f): Hz	Octave band range: Hz
31.5	22–44
63	44–88
125	88–177
250	177–354
500	354–707
1000	707–1414
2000	1414–2828
4000	2828–5657
8000	5657–11814

field of noise control are octave bands, that is a range or band of frequencies with the upper frequency limit f_u equal to twice the lower limit f_l. Each octave band is defined by the centre frequency f_{ob} where

$$f_{ob} = (f_u \times f_l)^{\frac{1}{2}} = (2f_1 \times f_1)^{1/2} = 1.414\, f_l \tag{7}$$

The commonly used octave bands are illustrated in *Table 5.1*.

Other bandwidths may be encountered in analysis and noise control work.

5.2.4 The decibel

The methods of assessing sound and developing noise criteria are complex and there are two approaches that may be used to obtain a quantitative measure of subjective noise. Either measure the physical characteristics of the sound over the frequency range on a meter and correct for the response of the ear, or alternatively, devise a measurement system which gives an output of loudness rather than amplitude. The first method can be undertaken but is rather long-winded 'by hand'. It can be approximated in instrumentation by feeding the sound pressure level into an electronic weighting network which automatically corrects the octave band levels and adds them together to give an approximation of the subjective level. The unit of measurement used is the decibel (dB).

The shapes of the weighting curves most commonly used are shown in *Figure 5.5* and the sound pressure level curve that follows the hearing characteristic of the ear is termed the A weighted curve. This curve gives the most widely used unit (dBA) for quantifying a noise level. Other curves on the weighting network are the B, C and D. Curves B and C are not commonly used and D is sometimes used for aircraft noise.

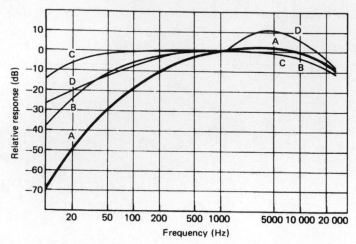

Figure 5.5 Weighting curves

5.2.5 Sound power level

The sound pressure level (L_p) must always be defined at a specific point in relation to the noise source. It is dependent on the location relative to the source, and the environment around the source and the receiver. The Sound Power Level (L_w) is a measure of the total acoustic energy radiated and is independent of the environment around the source. It is defined as:

$$L_w \text{ (dB)} = 10 \log_{10} \frac{W}{120} \tag{8}$$

where W is the power in watts. In many cases it is calculated from sound pressure levels by the approximate formula:

$$L_w = L_p + 10 \log_{10} A \qquad (\text{m}^2) \tag{9}$$

$$\text{or} \quad L_p + 10 \log_{10} A - 10 \qquad (\text{ft}^2) \tag{10}$$

where A = the surface area of measurement.

The concept of sound power level is best illustrated by the following analogy.

Suppose that the temperature measured at 1 metre from a 1 kW electric fire is the same as that measured 1 metre from a large steam boiler. Although the temperatures are the same, the boiler would be radiating more heat because it is radiating over a larger surface area. Similarly, the sound pressure level 1 metre from a large industrial cooling tower may be the same as that 1 metre from a small pump, but the acoustic power radiated by the cooling tower would be very much greater.

The sound power level is used to calculate the reverberant and pressure level and community noise.

5.3 Transmission of sound

The simplest concept in the transmission of noise is the inverse square law. For example, if the distance from the source is doubled, the reduction in noise level would be:

$$10 \log_{10} \left(\frac{2}{1}\right)^2 = 10 \log_{10} 4 = 6 \, \text{dB} \tag{11}$$

Where the sound power level is known the resulting sound pressure level at a distance is given by:

$$L_p = L_w - 10 \log 2\pi r^2, \text{ where } r \text{ is in metres} \tag{12}$$

Where distances exceed 200–300 metres other absorption factors should be taken into account. These include atmospheric attenuation and 'ground' absorption effects. Atmospheric attenuation (molecular absorption) is proportional to distance, but 'ground' absorption is dependent on the intervening terrain between the source and receiver.

5.4 The sound level meter

The majority of noise control work undertaken by the safety adviser will involve the measurement and possibly the analysis of noise. It is therefore important that the use and the limitations of sound pressure level measurements and sound level meters are understood. Errors in measurement technique or interpretation could lead to costly mistakes or over-specifying in remedial measures.

5.4.1 The instrument

There are many sound level meters on the market, but all work in a similar manner. The basic hand-held set (*Figure 5.6*) consists of a microphone, an amplifier with a weighting network and a read-out device in the form of a meter or digital presentation. The microphone converts the fluctuating sound pressure into a voltage which is amplified and weighted (A, B or Linear etc.). The electrical signal then drives a meter or digital read-out. In many instruments an octave or ⅓ octave band filter is incorporated, to enable a frequency analysis to be performed. The signal would then by-pass the weighting network and be fed into the read-out device via the frequency filters.

The difference between the two available grades of sound level meter, precision and industrial, is effectively the degree of accuracy of the

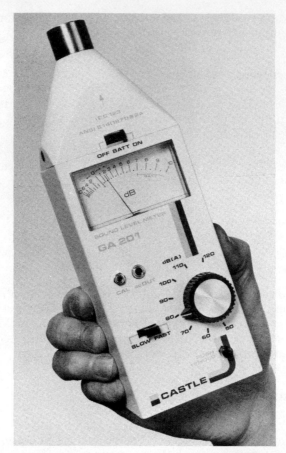

Figure 5.6 Industrial hand-held sound level meter. (Courtesy General Acoustics Ltd)

measurements particularly at high and low frequency. Meters should comply with British Standards[1] or with IEC[2] recommendations.

5.4.2 Use of a sound level meter

For the operation of sound level meters of different types the safety adviser should refer to the manufacturer's instruction book. The prime requirement for any instrument for noise measurement is that it should not be more sophisticated than necessary and it should be easy to use and calibrate. A typical sound level meter for use by the safety adviser should have the facility for measuring dBA and octave band sound pressure levels. More sophisticated meters have facilities for measuring equivalent noise level (L_{eq}, $L_{EP.d}$) (see section 5.6) and undertaking statistical analysis.

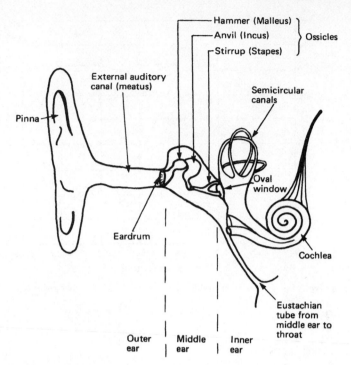

Figure 5.7 Diagram of human ear

5.5 The ear

5.5.1 Mechanism of hearing

Before appreciating how deafness occurs the manner in which the ear works should be understood. Briefly, the sound pressure waves enter the auditory canal (*Figure 5.7*) and cause the eardrum (tympanic membrane) to move in sympathy with the pressure fluctuations. Three minute bones called the hammer, anvil and stirrup (the ossicles) transmit the vibrations via the oval window to the fluid in the inner ear and hence to the minute hairs in the cochlea. These hairs are connected to nerve cells which respond to give the sensation of hearing. Damage to, or deterioration of, these hairs, whether by exposure to high noise levels, explosions or ageing will result in loss of hearing.

5.5.1.1 Audiogram

The performance of the ear is evaluated by taking an audiogram. Audiograms are normally performed in an Audiology Room which has a very low internal noise level and is vibration isolated. The audiogram is obtained by detecting the levels at which specific tones can be heard and cease to be heard by the person being tested.

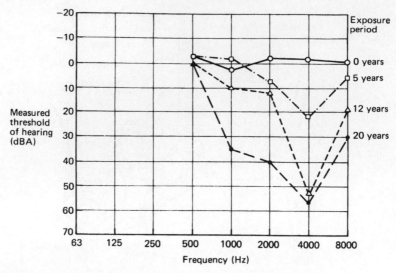

Figure 5.8 Typical audiograms showing progressive loss of hearing

5.5.1.2 Hearing loss

Noise-induced hearing loss (occupational hearing loss) is caused by over-stimulus of the receptor cells in the cochlea resulting in auditory fatigue. In its early stages it is often shown as an increase in the threshold of hearing – 'temporary threshold shift'. It may be accompanied by a ringing in the ears (tinnitus) which is indicative of temporary hearing damage. If exposure to high noise levels was continued, the result would be a 'permanent threshold shift' or noise-induced deafness. The effects of noise-induced deafness can be seen by the dip in the audiogram at 4 kHz, which deepens and widens with continuing exposure. Eventually the speech frequencies are affected (*Figure 5.8*), the sufferer loses the ability to hear consonants, and speech sounds like a series of vowels strung together.

Similar effects occur with age-induced hearing loss, which should always be considered when checking hearing impairment.

5.6 The equivalent noise level

In general the noise level in the community or inside a factory will vary with time. The equivalent noise level (L_{eq}) is defined as the notional steady noise level which, over a given period of time, would deliver the same amount of sound energy as the fluctuating level. Thus to maintain the L_{eq} when SPL is doubled, i.e. increased by 3 dB, exposure time must be halved (*Table 5.2*). The equivalent noise level concept forms the basis of the exposure criteria used in the Noise at Work Regulations 1989[3] which calls it 'daily personal noise exposure' ($L_{EP.d}$). Where the fluctuation is not

**Table 5.2 Exposure times vs. noise
levels for an equivalent noise level of
90 dBA**

Noise level (dBA)	Exposure time (h)
87	16
90	8
93	4
96	2
99	1
102	½
105	¼

well defined the calculations can be done electronically using a dosimeter
or statistical analyser[4]. Transient noises also require statistical analysis
and those measurements often required are:

L_{10} – the noise level exceeded for 10% of the time (average peak).
L_{50} – the noise level exceeded for 50% of the time (mean level).
L_{90} – the noise level exceeded for 90% of the time (average background
     level).

5.7 Community noise levels

Noise generated in a factory may affect not only those employed in the
factory but those living in the immediate neighbourhood, especially if
machinery is run during night-time. Legislation has been brought into
effect to give those affected the right of redress.

5.7.1 Control of Pollution Act 1974[5]

The Control of Pollution Act provides general legislation for limiting
community noise as well as other pollutants. No specific limits are set,
but the Act empowers the local authority to require a reduction in noise
emission and impose conditions for noisy operations, e.g. specify a level
of noise emission for a particular operation which must not be exceeded
at certain given times. In certain instances they may not only set the limits
but specify how they are to be met or how equipment is to be operated,
referring to the relevant British Standard[6].

5.7.2 Method of rating industrial noise in the neighbourhood

Criteria that can be used to assess the reasonableness of complaints and
for setting noise limits for design purposes are contained in a British

Standard[7] which in itself does not recommend specific limits, but predicts the likelihood of complaints. These predictions are based on measured or predicted noise levels, corrections for the noise characteristic and measured or notional background noise levels.

5.7.3 Assessing neighbourhood noise

The following procedure is typical for evaluating the existing neighbourhood noise situation:

1 Identify critical areas outside the factory by taking measurements over a reference period, as defined in the standard.
2 Note which equipment in the factory is operating during the measurement period to help evaluate the major sources.
3 Measure the L_{90} or background noise level. If the factory is always operating, the notional background noise level should be assessed using BS 4142[7]. Comparison of the factory noise level with the background level as explained in that British Standard, or with any limit set by the local authority will show whether further action is required.
4 Where there is possibility of complaints, or complaints have been made, a noise control programme should be introduced to reduce the noise from the major sources identified in stage (2).

5.8 Work area noise levels

Legislation, in the form of the Noise at Work Regulations 1989[3], has been enacted to limit the risk of hearing damage to workpeople while at work. These regulations give effect to an EC directive[8] on the protection of workers from the risks related to noise at work. Under the regulations, employers are required to take certain actions where noise exposure reaches 'Action Levels'

First action level   – $L_{EP.d}$ = 85 dB(A)
Second action level – $L_{EP.d}$ = 90 dB(A)
Peak action level   – a peak sound pressure of 200 pascals.

Where employees are likely to be exposed to noise levels at or above the first action or peak action level, the employer must ensure that a noise survey is carried out by a competent person (reg. 4). Records of this assessment must be kept (reg. 5). In some cases the level of noise to which an employee is exposed varies considerably over the duration of a shift making it difficult to assess the exposure from a single or a few spot measurements. The use of a dosimeter will enable a single figure to be derived for a shift. The dosimeter is worn with the microphone close to the ear for the period of the shift and the results either read directly from the instrument or downloaded to a computer. When using the dosimeter it is important that the wearer works to his or her normal pattern and does not spend longer than normal periods in a noisy environment.

Where a noise risk has been identified there is a requirement on the employer to reduce the emission to the lowest reasonably practicable level (regs 6 and 7).

If it is not reasonably practicable to reduce noise exposure levels to below 90 dB(A) $L_{EP.d}$ then the employer is required to provide personal hearing protection (reg. 8). An alternative to personal protection is the provision of hearing havens from which the worker can carry out his duties. Areas where there remains a hearing risk shall be designated 'ear protection zones' and be identified as such (reg. 9).

All hearing protection equipment must be kept in good order (reg. 10) and employees are required to use properly and look after any personal protective equipment issued to them. Considerable emphasis is placed on keeping employees informed (reg. 11) of:

- risks to their hearing,
- steps being taken to reduce the risk,
- procedures for obtaining personal protective equipment, and
- their obligations under the Regulations.

Regulation 12 modifies s. 6 of HSW to impose a duty on manufacturers and suppliers to inform customers when the level of noise emitted by their product exceeds the first action level, i.e. 85 dB(A) $L_{EP.d}$.

5.9 Noise control techniques

Before considering methods of noise control, it is important to remember that the noise at any point may be due to more than one source and that additionally it may be aggravated by noise reflected from walls (reverberant noise) as well as the noise radiated directly from the source. With any noise problems there are the three distinct elements shown in *Figure 5.9* – source, path and receiver.

Having identified the nature and magnitude of any noise problem, the essential elements of a noise control programme are provided. Where a problem is evident, there are three orders of priority for a solution:

1 Engineer the problem out by buying low noise equipment, altering the process or changing operating procedures.
2 Apply conventional methods of noise control such as enclosures or silencers.
3 Where neither of the above approaches can be used, the last resort of providing personal protection should be considered.

5.9.1 Source

Although the control of noise at source is the most obvious solution, the feasibility of this method is often limited by machine design, process or operating methods. While immediate benefits can be obtained, this method should be regarded as a long-term solution.

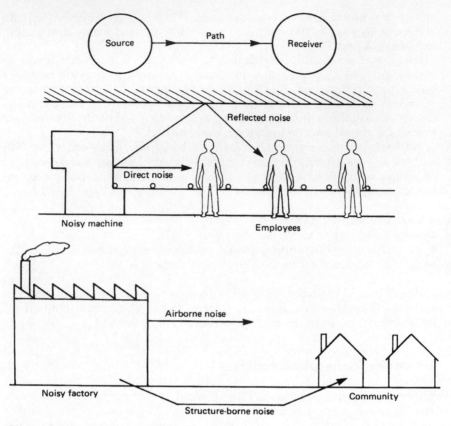

Figure 5.9 Noise source, path and receiver

5.9.2 Path

(a) Orientation and location
Control may be achieved by moving the source away from the noise
sensitive area. In other cases where the machine does not radiate equally
in all directions, turning it round can achieve significant reductions.

(b) Enclosure
Enclosures which give an attenuation of between 10 and 30 dBA[9] are the
most satisfactory solution since they will control both the direct field and
reverberant field noise components. In enclosing any source, the
provision of adequate ventilation, access and maintenance facilities must
be considered. A typical enclosure construction is shown in *Figure 5.10*.
The main features are an outer 'heavy' wall with an inner lining of an
acoustically absorbent material to minimise reverberant build-up inside
the enclosure. An inner mesh or perforate panel may be used to minimise
mechanical damage.

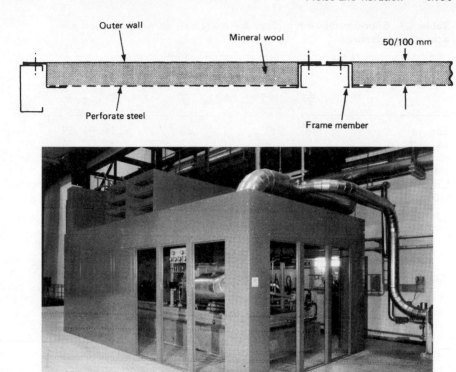

Figure 5.10 Noise enclosure. (Above) cross-section through typical noise enclosure wall; (below) noise enclosure with access doors removed. (Courtesy Ecomax Acoustics Ltd)

The sound reduction, attenuation or insertion loss is defined as the difference in sound pressure level or sound power level before and after the enclosure (or any other form of noise control) is installed. The performance of the enclosure will be largely dependent on the sound reduction index (SRI) of the outer wall[10], assuming approximately 50% of the internal surface is covered with mineral wool or other absorption materials[11]. Typical values of sound reduction index for materials used for enclosures are shown in *Table 5.3*. Absorption coefficients are shown in *Table 5.4*.

When considering an enclosure for any item of equipment, it is essential to consider a number of other aspects as well as the noise reduction required. Ventilation may be required to prevent overheating of the equipment being enclosed. Where ventilation is required each vent should be silenced. Special consideration should be given to the access requirements of maintenance and production, and the designers should ensure that these requirements are considered at an early design stage. In selecting any form of noise control, care should be taken to ensure that the

Table 5.3. Sound reduction indices for materials commonly used for acoustic enclosures

Material	Octave band centre frequency: Hz							
	63	125	250	500	1K	2K	4K	8K
22 g steel	8	12	17	22	25	26	25	29
16 g steel	12	14	21	27	32	37	43	44
Plasterboard	10	15	20	25	28	31	34	38
1/4 in plate glass	12	16	19	21	22	36	31	34
Unplastered brickwork (100 mm thick)	22	31	36	41	45	(50)	(50)	(50)

Bracketed figures refer to practical installations rather than test conditions.

Table 5.4 Typical sound absorption coefficients

Material	Octave band centre frequency: Hz							
	63	125	250	500	1K	2K	4K	8K
25 mm mineral wool	0.05	0.08	0.25	0.50	0.70	0.85	0.85	0.80
50 mm mineral wool	0.10	0.20	0.55	0.90	1.00	1.00	1.00	1.00
100 mm mineral wool	0.25	0.40	0.80	1.00	1.00	1.00	1.00	1.00
Typical perforated ceiling tile with 200 mm air space behind	0.30	0.50	0.80	0.95	1.00	1.00	1.00	1.00
Glass	0.12	0.10	0.07	0.04	0.03	0.02	0.02	0.03
Concrete	0.05	0.02	0.02	0.02	0.04	0.05	0.05	0.06

N.B. Concrete would normally be used for floor of enclosure.

equipment will physically withstand an industrial environment especially if it is particularly hostile. It must be robust and be capable of being dismantled and reassembled.

(c) Silencers[12]

Silencers are used to suppress the noise generated when air, gas or steam flow in pipes or ducts or are exhausted to atmosphere. They fall into two forms:

1 Absorptive, where sound is absorbed by an acoustical absorbent material.
2 Reactive, where noise is reflected by changes in geometrical shape.

Figure 5.11(a) shows a typical layout for an absorptive silencer, while *Figure 5.11(b)* shows a combination of the two types. The absorptive

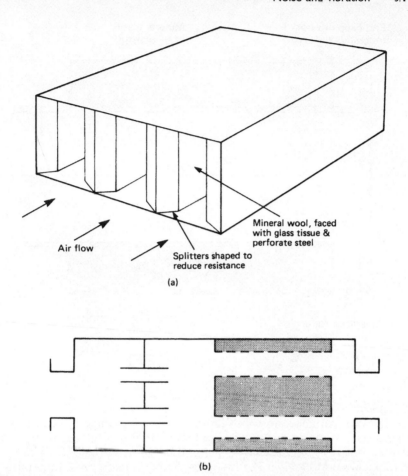

Mineral wool, faced
with glass tissue &
perforate steel

Air flow

Splitters shaped to
reduce resistance

(a)

(b)

Figure 5.11 Typical silencers (a) absorptive splitter silencer; (b) combination reactive/absorptive silencer

silencer normally has the better performance at higher frequencies, whereas the reactive type of silencer is more effective for controlling low frequencies.

The performance of splitter type of silencers is dependent on its physical dimensions. In general:

1 Sound reduction or insertion loss increases with length.
2 Low frequency performance increases with thicker splitters and reduced air gap.

Similarly for cylindrical silencers, the overall performance improves with length and the addition of a central pod. Performance would be limited by the sound reduction achievable by the silencer casing and other flanking paths. Typically no more than 40–50 dB at the middle frequencies could be expected without special precautions.

End cappings not
touching pipe wall

Mineral wool
or glass wool

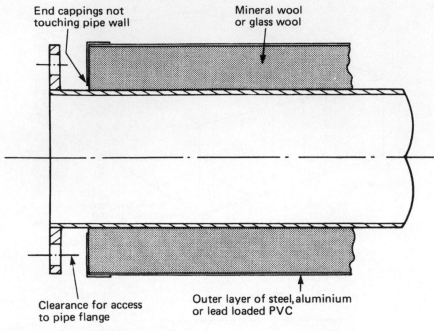

Clearance for access
to pipe flange

Outer layer of steel, aluminium
or lead loaded PVC

Figure 5.12 Pipe lagging

(d) Lagging[13]
On pipes carrying steam or hot fluids thermal lagging can be used as an alternative to enclosure and can achieve attenuations between 10 and 20 dBA, but it is only effective at frequencies above 500 Hz. The cross-section shown in *Figure 5.12* illustrates the main features of mineral wool wrapped around the pipes with an outer steel, aluminium or lead loaded vinyl layer. It is important that there is no contact between the outer layer and the pipe wall, otherwise the noise-reducing performance may be severely limited.

(e) Damping[14]
Where large panels are radiating noise a significant reduction can be achieved by fitting proprietary damping pads, fitting stiffening ribs or using a double skin construction.

(f) Screens
Acoustic screens (*Figure 5.13*) are effective in reducing the direct field component noise transmission by up to 15 dBA[15]. However, they are of maximum benefit at high frequencies, but of little effect at low frequencies and their effectiveness reduces with distance from the screen.

(g) Absorption treatment
In situations where there is a high degree of reflection of sound waves, i.e. the building is 'acoustically hard', the reverberant component can

Figure 5.13 Acoustic screens. (Courtesy Ecomax Acoustics Ltd)

Figure 5.14 Acoustic absorption treatment showing suspended panels and wall treatment. (Courtesy Ecomax Acoustics Ltd)

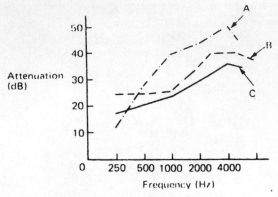

Figure 5.15 Attenuation characteristics of different hearing protectors (A: ear-muffs; B: plastic foam ear-plugs; C: acoustic wool ear-plugs)

dominate the noise field over a large part of the work area. The introduction of an acoustically absorbent material in the form of wall treatment and/or functional absorbers at ceiling height as shown in *Figure 5.14* will reduce the reverberant component by up to 10 dBA[11], but will not reduce the noise radiated directly by the source.

5.9.3 Personnel protection[16]

The two major methods of personnel protection are the provision of a quiet room or peace haven, and the wearing of ear-muffs or ear-plugs. The peace haven is similar in construction to an acoustic enclosure, and is used to keep the noise out. Ear-muffs or ear-plugs should be regarded only as the final resort to noise control. Their selection should be made with care having regard for the noise source, the environment and comfort of the wearer. Earplugs are only generally effective up to noise levels of 100–105 dBA while ear-muffs can provide protection at higher noise levels to meet a 90 dBA criterion, for noise received by the wearer. Comparative attenuation characteristics for various personal hearing protection devices are shown in *Figure 5.15*.

5.9.4 Effective noise reduction practices

A number of practical techniques can be used as part of normal day-to-day operational and maintenance procedures that will achieve significant reductions in the noise emitted, will cost nothing or very little to implement and can, additionally, give worthwhile savings in energy. Some of these techniques are listed below:

5.9.4.1 On plant

1 Tighten loose guards and panels.
2 Use anti-vibration mounts and flexible couplings.
3 Planned maintenance with programme for regular lubrication for both oil and grease.
4 Eliminate unnecessary compressed air and steam leaks, silence air exhausts.
5 Keep machinery properly adjusted to manufacturer's instructions.
6 Use damped or rubber lined containers for catching components.
7 Switch off plant not in use, especially fans.
8 Use rubber or plastic bushes in linkages, use plastic gears.
9 Resite equipment and design-in noise control.
10 Specify noise emission levels in orders, i.e. 85 dB(A) at 1 metre.
11 Check condition and performance of any installed noise control equipment.

5.9.4.2 Community noise

1 Keep doors and windows closed during anti-social hours.
2 If loading is necessary, carry out during day.
3 Vehicle manoeuvring – stop engine once in position.
4 Check condition and performance of any noise control equipment and silencers.
5 Turn exhaust outlets from fans away from nearby houses.
6 Carry out spot checks of noise levels at perimeter fence, both during working day and at other times.

5.9.5 Noise cancelling

Recent experiments to cancel the effects of noise electronically by emitting a signal that effectively flattens the noise wave shape have achieved some success in low frequency operation and situations where the receiver position and the source emission are well defined. However, the industrial application of this technique is still in its infancy.

5.10 Vibration

Vibration can cause problems to the human body, machines and structures, as well as producing high noise levels. Vibration can manifest itself as a particle displacement, velocity or acceleration. It is more commonly defined as an acceleration and may be measured using an accelerometer. There are many types of accelerometer and associated instrumentation available which can give an analogue or digital readout or can be fed into a computerised analysis system. As with sound, the vibration component would be measured at particular frequencies or over a band of frequencies.

5.10.1 Effect of vibration on the human body[17]

Generally, it is the lower frequency vibrations that give rise to physical discomfort. Low frequency vibration (3–6 Hz) can cause the diaphragm in the chest region to vibrate in sympathy giving rise to a feeling of nausea. This resonance phenomenon is often noticeable near to large slow speed diesel engines and occasionally ventilation systems. A similar resonance affecting the head, neck and shoulders is noticeable in the 20–30 Hz frequency region while the eyeball has a resonant frequency in the 60–90 Hz range.

The use of vibratory hand tools, such as chipping hammers and drills which operate at higher frequencies, can cause 'vibration induced white finger' (VWF). The vibration causes the blood vessels to contract and restrict the blood supply to the fingers creating an effect similar to the fingers being cold. Currently there are a number of cases where employees have instituted legal proceedings for VWF.

The effects of vibration on the human body will be dependent on the frequency, amplitude and exposure period and hence it is difficult to generalise on what they will be. However, it is worthwhile remembering that in addition to the physiological effects vibration can also have psychological effects such as loss of concentration.

5.10.2 Protection of persons from vibration

Where the source of the vibration cannot be removed, protection from whole body vibrations can be provided by placing the persons in a vibration isolated environment. This may be achieved by mounting a control room on vibration isolators in such areas as the steel industry, or simply having isolated seating such as on agricultural machinery.

For segmental vibration such as VWF consideration should be given to alternative methods of doing the job such as different tools.

5.10.3 Machinery vibration

For machines that vibrate badly, apart from the increased power used and damage to the machine and its supporting structure, the vibrations can travel through the structure of the building and be radiated as noise at distant points (structure-borne noise).

Where the balance of the moving parts of a machine cannot be improved, vibration transmission can be reduced by a number of methods[18] of which the most commonly used are:

1 Mount the machine on vibration isolators or dampers.
2 Install the machine on an inertia block with a damping sandwich between it and the building foundations.

The method chosen will depend on the size and weight of the machine to be treated, the frequency of the vibration to be controlled and the

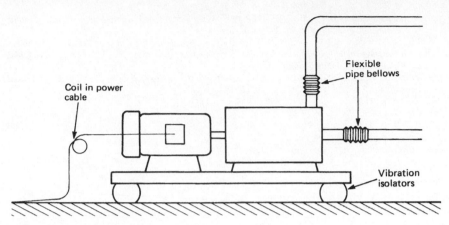

Figure 5.16 Vibration isolation

degree of isolation required. Whichever form of vibration isolation (*Figure 5.16*) is selected, care should be taken to ensure that the effect is not nullified by 'bridges'. For example, isolation of a reciprocating compressor set would be drastically reduced by rigid piping connecting it to its air receiver or distribution pipework, or by conduiting the cables to the motor. In severe cases rigid piping would fracture in a very short time.

5.11 Summary

The treatment of any noise source or combination of sources may use any of the control techniques individually or in combination. The selection of suitable measures will depend on:

1 The type of noise field – whether dominated by the direct noise radiated from the machine or the reverberant field.
2 The degree of attenuation required.
3 Whether work area limits or community noise limits are to be met.
4 Its cost effectiveness.

References

1. British Standards Institution, BS EN 60651:1994 *Specification for sound level meters*, BSI, London (1994)
2. International Electrotechnical Commission, IEC Standard 651, *Sound Level Meters*, IEC, Geneva (or British Standards Institution, London) (1979)
3. HM Government, *The Noise at Work Regulations 1989*, SI 1989 No. 1790, HMSO, London (1989)
4. Hassall, J. R. and Zaveri, K., *Acoustic Noise Measurements*, Bruel and Kjaer, Naerum, Denmark (1979)
5. *Control of Pollution Act 1974*, part III, HMSO, London (1974)
6. British Standards Institution, BS 5228:1975, *Code of practice for noise control on construction and demolition sites*, BSI, London (1975)

7. British Standards Institution, BS 4142:1997, *Method of rating industrial noise affecting mixed residential and industrial areas*, BSI, London (1997)
8. European Economic Community, Directive No. 86/188/EEC *On the protection of workers from the risks related to exposure to noise at work*, Official Journal No. L137 of 24 May 1986, p. 28, HMSO, London (1986)
9. Warring, R. A. (Ed.), *Handbook of Noise and Vibration Control*, 509, Trade and Technical Press, London (1970)
10. Ref. 9, p. 474
11. Ref. 9, p. 460
12. Ref. 9, p. 543
13. Ref. 9, p. 249
14. Ref. 9, p. 595
15. Ref. 9, p. 504
16. Ref. 9, p. 571
17. Ref. 9, p. 112
18. Ref. 9, p. 586

Further reading

Blitz, J., *Elements of Acoustics*, Butterworth, London (1964)
Beranek, L.L., *Noise and Vibration Control*, McGraw-Hill, New York (1971)
Health and Safety Executive, Health and Safety at Work Series Booklet No. 25, *Noise and the Worker*, HMSO, London (1976)
Health and Safety Executive, Report by the Industrial Advisory Subcommittee on Noise, *Framing Noise Legislation*, HSE Books, Sudbury (undated)
Sharland, I., *Wood's Practical Guide to Noise Control*, Woods of Colchester Ltd, England (1972)
Taylor, R., *Noise*, Penguin Books, London (1970)
Webb, J.D. (Ed.), *Noise Control in Industry*, Sound Research Laboratories Ltd, Suffolk (1976)
Burns, W. and Robinson, D., *Hearing and Noise in Industry*, HMSO, London (1970)
Burns, W. and Robinson, D., *Noise Control in Mechanical Services*, Sound Attenuators Ltd, and Sound Research Laboratories Ltd, Colchester (1972)
Health and Safety Executive, *Noise 1990 and You*, HSE Books, Sudbury (1989)

Chapter 6

Workplace pollution, heat and ventilation

F. S. Gill

The solution of many workplace environmental problems, whether due to the presence of airborne pollutants such as dust, gases or vapours, or due to an uncomfortable or stressful thermal environment, lies in the field of ventilation. Ventilation can be employed in three ways:

1 By using extraction as close to the source of pollution as possible to minimise the escape of the pollutant into the atmosphere. The extraction devices can be either hoods, slots, enclosures or fume cupboards coupled to a system of ducts, fans and air cleaners.
2 By providing sufficient dilution ventilation to reduce the concentration of the pollutants to what is thought to be a safe level.
3 By using air as a vehicle for conveying heat or cooling to a workplace to maintain reasonably comfortable conditions by employing air conditioning or a warm air ventilation system.

A flow of air which may be part of an industrial process can have a substantial effect upon the safety of the workplace by removing – or not – excessive heat, fume or dust. Such processes as ink drying, solvent collection, particulate conveying could fall into this category.

Before embarking upon the design for a ventilation system, it is necessary to assess the extent of the problem, that is, the amount of airborne pollution to be encountered in a workplace, and/or the degree of discomfort or stress expected from a thermal environment. Measurement and analysis techniques need to be devised and criteria and standards applied to the environment under consideration. Where measurement and analysis are concerned, the physics and the chemistry of the properties of the pollutant and its mode of emission need to be studied in such a way that a reliable and accurate assessment of the exposure of a worker can be made. As far as criteria and standards are concerned, medical evidence, biological research and epidemiological methods need to be applied to establish the relationship between the exposure and the long- and short-term effect upon the human body of the worker taking into account the duration of exposure and the work rate. It can be seen, therefore, that many scientific skills require to be involved before a judgement can be made and a method of control devised.

6.1 Methods of assessment of workplace air pollution

Airborne pollutants can be divided roughly into three groups:

1 dusts and fibres;
2 gases and vapours;
3 micro-organisms (bacteria, fungi etc.);

although some emissions from workplaces, for example oil mist from machine tools, could contain material from each group.

There is a wide range of techniques available to measure the degree of workplace pollution, some of which are described below. First it is important to decide what information is required as the technique chosen will determine whether the concentration of airborne pollution is measured:

(a) in the general body of the workroom at an instant of time or averaged over the period of work, the latter being known as a time weighted average (TWA); or
(b) in the breathing zone of a worker averaged over the period of work; or
(c) in the case of dusts, as the total or respirable airborne dust (respirable dust is that which reaches the inner part of the lungs).

If a time weighted average is required, it might also be necessary to know whether any dangerous peaks of concentration occurred during the work period.

Owing to air currents in workrooms, pollutants can move about in clouds; thus concentrations vary with place and time and some statistical approach to measurement may be required. Also some workers move about from place to place in and out of polluted atmospheres and, while workplace concentrations might be high, operator exposure levels might be lower.

In addition, workers may be exposed to a variety of different pollutants during the course of a shift, some being more toxic than others, so that the degree of exposure to each might be required to be known. When a mixture of pollutants occurs, the presence of one may interfere with the measurement of another. Therefore, the measurement of the degree of exposure of a worker must be undertaken with care. Occupational (or industrial) hygienists are trained in the necessary skilled techniques and should be called in to carry out the measurements.

6.1.1 Airborne dust measurement

The commonest method for measuring airborne dust is the filter method where a known volume of air is drawn through a pre-weighed filter paper (*Figure 6.1*) or membrane by means of a pump (*Figure 6.2*). The filter can be part of a static sampler located at a suitable place in the workroom or it can be contained in a special holder attached to a

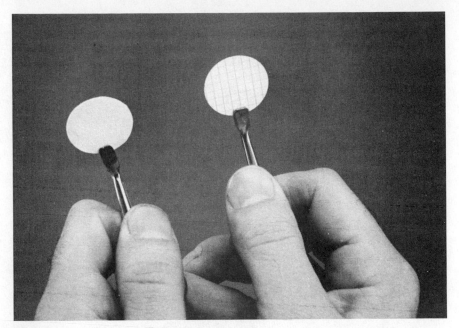

Figure 6.1 Dust sampling filters

person as close to the face as possible, usually fixed to the lapel or shoulder strap of a harness and connected by tubing to the pump which is attached to the wearer's belt. At the end of the sampling period the filter is weighed again and the difference in weight represents the weight of dust collected. This, divided by the total volume of air which has passed through the filter, gives the average concentration of dust over the period.

If a membrane-type filter made of cellulose acetate is used, it can be made transparent by the application of a clearing fluid, allowing the dust to be examined under a microscope and the particles or fibres counted if required. This is particularly important if fibrous matter such as asbestos is present.

Other filters can be chemically digested so that the residues can be further examined by a variety of chemical means. Types of filter are available that are suitable for examining the collected dust by X-ray diffraction, X-ray fluorescent techniques or by a scanning electron microscope. Thus, the choice of filter must be related to the type of analysis required. Weighing must be accurate to five decimal places of grams and, as air humidity can affect the weight of some filters, preconditioning may be required.

When on personal samplers the level of respirable dust is required, a device such as a cyclone is used to remove particles above 10 μm in diameter. Static samplers use a parallel plate elutriator for this purpose which allows the larger particles to settle so that only the respirable dust reaches the filter. With all separation devices the airflow rate must be

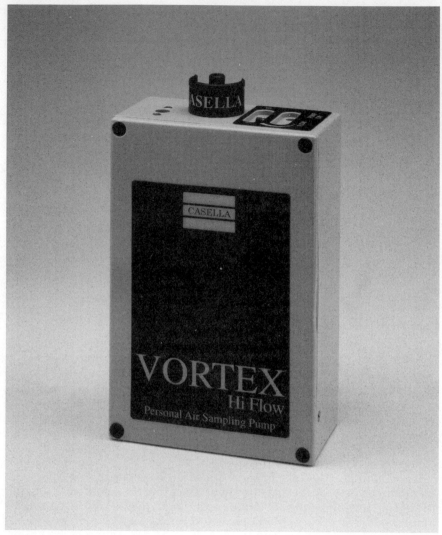

Figure 6.2 Personal dust sampling pump. (Courtesy Casella Ltd)

controlled within close limits to ensure that the correct size fraction separation occurs.

Other techniques for static dust measurement include those which use the principle of measuring the amount of light scattered by the dust and one that uses a technique which measures the oscillation of a vibratory sensor which changes in frequency with the amount of dust deposited in it; another uses the principle of absorption of beta-radiation by the amount of dust deposited on a thin polyester film.

Further details of dust sampling and measurement techniques are given in references 1, 2 and 4.

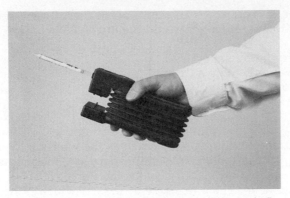

Figure 6.3 Draeger bellows pump and tube. (Courtesy Draeger Ltd)

6.1.2 Airborne gases and vapours measurement

The number of techniques available in this field is vast and the range is almost as wide as chemistry itself. Instruments can be used which are specific to one or two gases while others use the principle of infrared absorption and can be tuned to be sensitive to a range of selected gases. The principle of change of colour of paper or crystals is also used for specific gases and vapours; detector tube and impregnated paper samplers are of this type, but difficulties can be experienced if more than one gas is present as one may interfere with the detection of the other (*Figure 6.3*).

Techniques for unknown pollutants involve collecting a sample over a period of time and returning it to the laboratory for detailed chemical analysis. Collection can be in containers such as bladders, bags, cylinders, bottles, syringes or on chemically absorbent materials such as activated charcoal or silica gel. Those methods which use containers may lose some of the collected gases or vapours by adherence to the inside of the container. The chemical absorbers are usually contained in small glass or metal tubes connected to a low-volume sampling pump and can be worn by a worker in a similar fashion to the personal dust samplers.

All these techniques require a good knowledge of chemistry if reliable results are to be obtained as many problems exist in the collection and analysis of gases and vapours[2,3,4].

6.2 Measurement of the thermal environment

Many working environments are uncomfortable owing to excessive heat or cold in one form or another; some can be so extreme that they lead to heat or cold stress occurring in the workers. When investigating these environments, it is important to take into account the rate of work and the type of clothing of the worker as these affect the amount of heat the body is producing and losing.

To obtain a correct assessment of a thermal environment, four parameters require to be measured together:

1 The air dry bulb temperature.
2 The air wet bulb temperature.
3 The radiant temperature.
4 The air velocity.

If any one of these is omitted, then an incomplete view is obtained. The sling psychrometer (sometimes known as the whirling hygrometer) will measure wet and dry bulb temperatures, a globe thermometer responds to radiant heat, and air velocity can be measured by an airflow meter or a katathermometer. There are several indices which bring together the four measurements and express them as a single value: some also take into account work rate and clothing. A number of these indices are listed below, and their values can be calculated or derived from charts.

1 The Wet Bulb Globe Temperature index (WBGT) can be calculated from the formula:

   For indoor environments
   WBGT = 0.7 × natural wet bulb temperature + 0.3 × globe
          temperature

   For outdoor environments
   WBGT = 0.7 × natural wet bulb temperature + 0.2 × globe
          temperature + 0.1 × dry bulb temperature

2 Corrected Effective Temperature index (CET) can be obtained from a chart and can take into account work rate and clothing.
3 Heat Stress Index (HSI) can be calculated or obtained from charts and takes into account clothing and work rate, and from it can be obtained recommended durations of work and rest periods.
4 Predicted four hour sweat rate (P4SR) can be obtained from charts and takes into account work rate and clothing.
5 Wind chill index, as its name suggests, refers to the cold environment and uses only dry bulb temperature and air velocity but takes into account the cooling effect of the wind.

These five indices are considered in detail, including charts and formulae, in references 5 and 6.

6.3 Standards for workplace environments

Authorities from several countries publish recommended standards for airborne gases, vapours, dusts, fibre and fume. For many years the main players in the standard setting field have been the United States of America (US) through the American Conference of Governmental

Industrial Hygienists (ACGIH) who publish threshold limit values (TLVs)[8] and the United Kingdom (UK) through the Health and Safety Executive (HSE) who publish Occupational Exposure Limits in their Guidance Note Series EH40[7]. Both these sources update annually. However, recently another important player has emerged and that is the European Union (EU). The Directorate General of the European Commission has set up a Scientific Experts Group whose task is to develop a list of European occupational exposure limits[14] for member states to consider when setting their own standards.

The occupational exposure limits for the UK are published in EH40. Essentially these standards are in two parts: maximum exposure limits (MELs) and occupational exposure standards (OESs), and they have a legal status under regulation 7 of COSHH[9].

6.3.1 Maximum Exposure Limits

Regulation 7(6) of COSHH requires that where an MEL is specified the control of exposure shall be such that the level of exposure is reduced as low as reasonably practicable and in any case below the MEL. Where short-term exposure limits are quoted they shall not be exceeded. In the 1998 edition of EH40 there are 59 MELs quoted.

6.3.2 Occupational Exposure Standards

Regulation 7(7) of COSHH requires that where an OES is quoted the control of exposure shall be treated as adequate if the OES is not exceeded or, if exceeded, the employer identifies the reasons and takes appropriate action to remedy the situation as soon as is reasonably practicable. There are some 700 OESs published.

6.3.3 Units and recommendations

The standards in the above document are quoted in units of parts per million (ppm) and milligrams per cubic metre (mg m^{-3}). They are given for two periods: Long-term exposure which is an 8-hour time weighted average (TWA) value and a short-term exposure which is a 15-minute TWA. Certain substances are marked with a note 'Sk' which indicates that they can also be absorbed into the body through the skin, others have a 'Sen' notation indicating that they may possibly sensitise an exposed person.

It should be remembered that all exposure limits refer to healthy adults working at normal rates over normal shift duration. In practice it is advisable to work well below the recommended value, as low as one-quarter, to provide 'a good margin of safety'.

6.4 Ventilation control of a workplace environment

As a result of the COSHH regulations there is a legal duty to control substances that are hazardous to health. The Approved Code of Practice (ACOP)[9] associated with these regulations sets out in order the methods that should be used to achieve adequate control. Extract and dilution ventilation are two of the methods mentioned. These regulations also require the measurement of the performance of any ventilation systems that control substances that are hazardous to health. The places where measurements are required to be taken are listed in para. 61 of the ACOP.

6.4.1 Extract ventilation

In the design of extract ventilation it is important to create, at the point of release of the pollutants, an air velocity sufficiently strong to capture and draw the pollutants into the ducting. This is known as the capture velocity and can be as low as 0.25 m/s for pollutants released gently into still air such as the vapour from a degreasing tank or as much as 10 m/s or more for heavy particles released at a high velocity from a device such as a grinding wheel. The capturing device can be a hood, a slot or an enclosure to suit the layout of the workplace and the nature of the work but the more enclosure that is provided and the closer to the point of emission it is placed, the more effective will be the capture.

Difficulty can be experienced with moving sources of pollution such as the particles from hand-held power saws and grinders. In these circumstances high velocity low volume extractors can be fitted to the tools using flexible tubing of 25–50 mm diameter to draw the particle-laden air to a cleaner which contains a high efficiency filter and a strong suction fan (*Figure 6.4*).

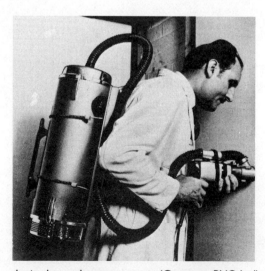

Figure 6.4 High velocity low volume extractor. (Courtesy BVC Ltd)

Figure 6.5 Portable collecting hood. (Courtesy Myson Marketing Services Ltd)

Hoods attached to larger diameter flexible tubing can be used for extraction from the larger moving sources such as welding over wide areas, but owing to the higher weight of these devices some form of movable support system is required (*Figure 6.5*).

When siting a capture hood or slot, advantage should be taken of the natural movement of the pollutants as they are released. For example, hot substances and gases are lighter than air and tend to rise, thus overhead capture might be most suitable, whereas some solvent vapours when in concentrated form are heavier than air and tend to roll along horizontal surfaces, so capture points are best placed at the side. Care must be taken to ensure that all contaminants are drawn away from the breathing zone of the worker – this particularly applies to places where workers have to lean over or get close to their work. It is important to note that whenever extract ventilation is exhausted outside, a suitably heated supply of make-up air must be provided to replace that volume of air discarded.

There are established criteria for the design of extract systems[10].

6.4.2 Dilution ventilation

This method of ventilation is suitable for pollutants that are non-toxic and are released gently at low concentrations and should be resorted to only if it is impossible to fit an extractor to the work station. It should not be used if the pollutants are released in a pulsating or intermittent way or if they are toxic. The volume flow rate of air required to be provided must be calculated taking into account the volume of the pollutants released, the concentration permitted in the workplace and a factor of safety which allows for the layout of the room, the airflow patterns created by the ventilation system, the toxicity of the pollutant and the steadiness of its release[11,12].

Hourly air change rates are sometimes quoted to provide a degree of dilution ventilation. The volume flow rate of air in cubic metres per hour is calculated by multiplying the volume of the room in cubic metres by the number of air changes recommended. There are recommended air change rates for a range of situations[13].

6.5 Assessment of performance of ventilation systems

In addition to the testing of the airborne concentrations of pollutants, it is necessary, and indeed is a requirement of COSHH, to check airflows and pressures created in a ventilation system to ensure that it is working to its designed performance by measuring:

1 Capture velocity.
2 Air volume flow rates in various places in the system.
3 The pressure losses across filters and other fittings and the pressures developed by fans.

The design value of these items should be specified by the maker of the equipment. Therefore, instruments and devices are required to:

1 Trace and visualise airflow patterns.
2 Measure air velocities in various places.
3 Measure air pressure differences.

Figure 6.6 Smoke tube

Figure 6.7 Vane anemometer. (Courtesy Air Flow Developments Ltd)

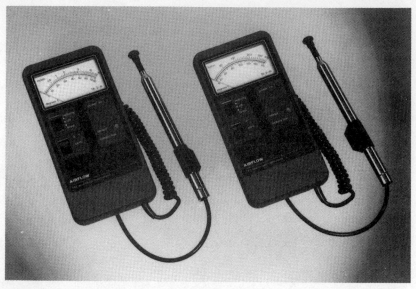

Figure 6.8 Heated head air meter. (Courtesy Airflow Developments Ltd)

Air flow patterns can be shown by tracers from 'smoke tubes' which produce a plume of smoke when air is 'puffed' through them (*Figure 6.6*). For workplaces where airborne particles are released it is possible to visualise the movement of the particles by use of a dust lamp. This shines a strong parallel beam of light through the dust cloud highlighting the particles in the same way that the sun's rays do in a darkened room.

Air velocities can be measured by a variety of instruments but vane anemometers and heated head (hot wire or thermistor) air meters are the most common. Vane anemometers (*Figure 6.7*) have a rotating 'windmill' type head coupled to a meter and are most suitable for use in open areas such as large hoods and tunnels. The heated head type of air meter (*Figure 6.8*) is more suitable for inserting into ducting and small slots and is more versatile than the vane anemometers except that it is unsuitable for use in areas where flammable gases and vapours are released. Most air flow measuring instruments require checking and calibration from time to time.

One instrument which requires no calibration but is only effective in measuring velocities above approximately 3 m/s is the pitot-static tube which, in conjunction with a suitable pressure gauge, measures the velocity component of the pressure of the moving air which can be converted to air velocity by means of the simple formula:

$$p_v = \tfrac{1}{2}\rho v^2 \quad \text{or} \quad v = \sqrt{\frac{2p_v}{\rho}}$$

where p_v = velocity pressure (N/m^2 or Pa); ρ = air density (usually taken to be 1.2 kg/m^3 for most ventilation situations); and v = air velocity (m/s).

Pitot-static tubes are small in diameter and can easily be inserted into ducting.

All the above air velocity measuring instruments need to be placed carefully in an airstream so that their axes are parallel to the stream lines; any deviation from this will give errors.

Differences in air pressure can be measured by a manometer or U-tube gauges filled with water or paraffin, placed either vertically or, for greater accuracy, inclined. If the two limbs of the gauge are coupled by flexible plastic or rubber tubing to either side of the place to be measured, such as a fan or a filter, then the difference in height between the two columns of the tube indicates the pressure difference. Pressure tappings in ductings must be at right angles to the air flow to measure what is termed 'static pressure'.

Liquid-filled gauges are prone to spills and the inclusion of bubbles and before use must be carefully levelled and zeroed. Diaphragm pressure gauges avoid these problems but need to be checked for accuracy from time to time. Electronic pressure gauges are also available.

Airflow measuring techniques vary to suit the application[4].

References

1. Brief, R. S., *Basic Industrial Hygiene*, Section 9, Exxon Corporation (1975)
2. ACGIH, *Air Sampling Instruments*, 8th edn, American Conference of Governmental Industrial Hygienists, Cincinnati, Ohio (1995)
3. Thain, W., *Monitoring Toxic Gases in the Atmosphere for Hygiene and Pollution Control*, Pergamon, Oxford (1980)
4. Gill, F.S. and Ashton, I., *Monitoring for Health Hazards at Work*, Chapter 4, 'Ventilation', Blackwell Science, Oxford (1998)
5. Youle, A., 'The thermal environment' chapter in *Occupational Hygiene* (Eds Harrington, J.M. and Gardiner, J., Blackwell Science, Oxford (1995)
6. Harrington, J.M., Gill, F.S., Aw, T.C. and Gardiner, K., *Occupational Health Pocket Consultant*, Blackwell Science, Oxford (1998)
7. Health and Safety Executive, Guidance Note EH40, *Occupational Exposure Limits*, HSE Books, Sudbury (1998)
8. ACGIH, *Threshold Limit Values for Chemical Substances and Physical Agents in the Workroom Environment*, American Conference of Governmental Industrial Hygienists, Cincinnati, Ohio (1998)
9. Health and Safety Executive, Legal series booklet no. L 5, *General COSHH ACOP (Control of substances hazardous to health), Carcinogens ACOP (Control of carcinogenic substances) and Biological agents (Control of biological agents). Control of Substances Hazardous to Health Regulations 1994. Approved Code of Practice*, HSE Books, Sudbury (1995)
10. British Occupational Hygiene Society, Technical Guide No. 7, *Controlling Airborne Contaminants in the Workplace*, Science Reviews Ltd, Leeds (1987)
11. Gill, F.S., 'Ventilation' chapter in *Occupational Hygiene* (Eds Harrington, J.M. and Gardiner, K), Blackwell Scientific, Oxford (1995)
12. ACGIH, *Industrial Ventilation*, 22nd edn, American Conference of Governmental Industrial Hygienists, Cincinnati, Ohio (1995)
13. Daly, B. B., *Woods Practical Guide to Fan Engineering*, chapter 2, Woods of Colchester Ltd (1978)

14. EEC Council Regulation no. EEC/793/93 *on the evaluation and control of the risks of existing substances*, EC, Luxembourg (1993)

Further reading

Ashton, I. and Gill, F.S., *Monitoring for Health Hazards at Work*, Blackwell Science, Oxford (1998)

Chapter 7

Lighting

E. G. Hooper and updated by Jonathan David

7.1 Introduction

Lighting plays an important role in health and safety, and lighting requirements are increasingly being included in legislation and standards, albeit that primary legislation tends to specify that lighting shall be 'sufficient and suitable'. Legislation whose content has lighting in its requirements includes that for the workplace[1], work equipment[2], docks[3], the use of electricity[4] and display screen equipment[5]. Most people prefer to work in daylight making the best possible use of natural light, though this may not always be the most energy efficient approach. However, for many working environments natural light is often insufficient for the whole working day, and in deeper spaces may not be adequate at any time. It therefore has to be supplemented or replaced by artificial lighting, usually electric lighting. The quality of the lighting installation can have a significant effect on health, productivity and the pleasantness of interior spaces in addition to its role in safety.

7.2 The eye

The front of the eye comprises, in simple terms, a lens to control the focusing point within the eye and an iris to control the light entering the eye. The back of the eye contains the retina which is made up of rod and cone shaped cells which are sensitive to light and are linked by optic nerves to the brain. The lens ensures that the image being viewed is focused on the retina and the iris controls the amount of light. Different cells in the retina are sensitive to different colours, and while the central part of the retina, known as the fovea, is sensitive to colours the peripheral areas are sensitive only to light intensity. A result is that colour vision disappears at low light levels.

7.3 Eye conditions

The eye is a very delicate and sensitive structure and is subject to a number of disorders and injuries requiring skilled treatment: some of these disorders are mentioned briefly below.

Conjunctivitis is an inflamed condition of the conjunctiva (the mucous membrane covering the eyeball) caused by exposure to dust and fume and occasionally to micro-organisms.

Eye strain, so called, is caused by subjecting the eye to excessively bright light or glare; the term is also used colloquially to describe the symptoms of uncorrected refractive errors. There is no evidence that the eye can be 'strained' simply by being used normally.

Accommodation is a term for the ability of the eye to alter its refractive powers and to adjust for near or distant vision. As the eye ages the lens loses its elasticity and hence its accommodation, thus affecting the ability to read and requiring corrective spectacles. In addition to this ageing process defects in accommodation can occur early in life, such as by the presence of conditions known as

1 *astigmatism* due to the cornea of the eye being unequally curved and affecting focus;
2 *hypermetropia*, or long sight, in which the eyeball is too short; and
3 *myopia*, or short sight, in which the eyeball is too long.

These defects can usually be corrected by spectacles.

Nystagmus is an involuntary lateral or up and down oscillating and flickering movement of the eyeball, and is a symptom of the nervous system observed in such occupations as mining.

Double vision is the inability of both eyes to focus in a co-ordinated way on an object usually caused by some defect in the eye muscles. It can be due to a specific eye injury, to tiredness or be a symptom of some illness. It may be a momentary phenomenon or may last for longer periods.

Colour blindness is a common disorder where it is difficult to distinguish between certain colours. The most common defect is red/green blindness and may be of a minor character where red merely loses some of its brilliance, or of a more serious kind where bright greens and reds appear as one and the same colour – a dangerous condition in occupations requiring the ability to react to green and red signals or to respond to colour coding of pipework or electrical cables.

Temporary blindness may be due to some illness but it can occur in the following circumstances:

1 Involuntary closure of the eyelids due to glare.
2 Impairment of vision due to exposure to rapid changes in light intensity and to poor dark adaptation or to excessively high light levels.

The act of seeing requires some human effort which is related to the environmental conditions. Even with good eyesight a person will find it difficult to see properly if the illumination (level of lighting) is not adequate for the task involved, e.g. for the reading of small print or working to fine detail. But no standard of lighting, however well planned, can correct defective vision and anyone with suspected visual disability should be encouraged to undergo an eye test and, if advised, wear corrective spectacles. Legislation now requires that employees working with visual display terminals (vdts) be offered free eye tests by their employers if they so request[5].

7.4 Definitions[6]

The following terms are used in connection with illumination:

Candela (cd) is the SI unit of luminous intensity, i.e. the measure describing the power of a light source to emit light.

Lumen (lm) is the unit of luminous flux used to describe the quantity of light emitted by a source or received by a surface.

Illuminance (symbol E, unit *lux*) is the luminous flux density of a surface, i.e. the amount of light falling on a unit area of a surface, 1 lux = $1 \, lm/m^2$.

Maintained illuminance is the average illuminance over the reference surface at the time maintenance has to be carried out. It is the level below which the illuminance should not drop at any time in the life of the installation.

Luminance (symbol L, unit cd/m^2) is the physical measure of the stimulus which produces the subjective sensation of brightness, measured by the luminous intensity of the light emitted or reflected in a given direction from a surface element divided by the projected area of the element in the same direction.

Luminance = (illuminance × reflection factor)/π

Brightness is the subjective response to luminance in the field of view dependent on the adaptation of the eye.

Reflectance factor is the ratio of the luminous flux reflected from a surface to the luminous flux incident upon it.

Incandescent lamp is a lamp where the passage of a current through a filament (usually coiled) raises its temperature to white heat (incandescence), giving out light. Oxidisation within the glass bulb is slowed down by the presence of an inert gas or vacuum sealing of the bulb. The most commonly used lamp is the General Service Lamp, but there also exists a wide range of decorative lamps. Higher efficiency incandescent lamps can be created by including in the bulb a small amount of a halogen element such as iodine or bromine. In such lamps, usually known as tungsten-halogen lamps, the halogen combines with the tungsten and is deposited on the inside of the bulb. When this compound approaches the filament it decomposes, owing to the high temperature, and deposits the tungsten back on the filament.

The European Commission is developing a scheme for energy rating of lamps commonly used for domestic purposes, along the lines of that for domestic appliances. It does not appear that it will apply to other lamp types or lamps sold to commercial and industrial organisations.

Electric discharge lamp is a lamp where an arc is created between two electrodes within a sealed and partially evacuated transparent tube. Depending on the format of the tube, the remaining gas pressure and the trace elements that are introduced, numerous different types of lamp can be produced:

1 *Low pressure sodium lamp* used chiefly for road lighting which produces a monochromatic yellow light but is highly efficient.
2 *Low pressure mercury lamp* – the ubiquitous 'fluorescent tube' in which the ultraviolet radiation from the discharge is converted to visible light by means of a fluorescent coating (phosphor) on the inside of the tube. Fluorescent lamps come in various forms:
    (a) Linear lamps, both full size (600–2400 mm long) and miniature (less than 600 mm long), come in a range of wattages and efficiencies as well as a range of whites and colours. Traditionally, while halophosphate phosphors were used, there was a trade-off between colour quality and efficiency; with modern triphosphor and multi-band lamps this is no longer the case.
    (b) Compact lamps, in both retrofit designs intended for existing installations and for newer installations when compatibility with other lamp types does not matter, come in a variety of formats and ratings from 5 W to 55 W.
3 *High pressure mercury lamp* is a largely obsolete type of lamp where light is produced by means of a discharge within an arc tube doped with mercury. The light tends to be bluish in colour and efficiency is lower than other currently used types of discharge lamp. It is still popular in some tropical countries because of its 'cool' light.
4 *High pressure sodium lamp* is similar to a mercury lamp except that the arc tube is doped with sodium giving a yellow light whose colour rendering and whiteness depend on the vapour pressure within the tube.
5 *Metal halide lamp* is similar to the mercury lamp except that the mercury is replaced by a carefully designed cocktail of rare earth elements. Colour rendering can be very good and efficiency is high with additional coloured light being generated by the suitable choice of elements in the cocktail. The small arc tube means that light control can be very good.

Induction lamp in which the lamp itself is simply a glass tube containing an inert gas and coated on the inside with a phosphor to convert the ultraviolet radiation to visible light. The discharge which takes place in the tube is initiated by an electric or microwave field outside the lamp by equipment containing a powerful electromagnet or a magnetron. Different manufacturers have adopted different physical formats. Efficiency is fairly high and, because there are no moving parts in the tube, lamp life can be extremely long making the lamp ideal where maintenance access is difficult.

Luminaire is a general term for all the apparatus necessary to provide a lighting effect. It usually includes all components for the mounting and protection of lamps, controlling the light distribution and connecting them to the power supply, i.e. the whole lighting fitting. Occasionally part of the control gear may be mounted remote from the luminaire.

7.5 Types of lighting

The selection of the source of light appropriate to the circumstances depends on several factors. It is important to consider efficiency, ease of installation, costs of installation and running, maintenance, lamp life characteristics, size, robustness and heat and colour output. The efficiency of any lamp (often termed efficacy) can be expressed in terms of light output per unit of electricity used (lumens per watt). Generally speaking, incandescent lamps are less efficient than discharge sources.

Type of lamp	Lumens per watt
Incandescent lamps	About 15
Tungsten halogen	Up to 22
High pressure sodium	Up to 140
Metal halide	Up to 100
Fluorescent	Up to 100
Compact fluorescent	Up to 85
Induction	Up to 65
Low pressure sodium	Up to 200

Note that smaller ratings are usually less efficient than larger ratings and that the above figures do not include losses within the control gear needed for all but incandescent lamps. Note also that control gear losses can differ markedly between brands. A rating scheme for efficiency of ballasts for fluorescent lamps has been introduced in Europe[7].

In any choice between incandescent and the other types of lamp the total lighting costs must take into account not only running costs but also installation and replacement costs. Incandescent lamps are much cheaper to buy and install, they give out light immediately they are switched on and they can be dimmed easily, but they are more expensive to run and have short lives, thus increasing maintenance costs. High pressure discharge and fluorescent lamps cost more to install but their greater efficiency and longer lives make them more cost effective for general lighting. Linear and compact fluorescent lamps come to full light output reasonably quickly but discharge lamps need some time to strike and then achieve maximum light output, and may need several minutes to cool before they will restrike if accidentally extinguished. Hot restrike is possible for some lamps but is expensive.

In larger places of work the choice is often between discharge and fluorescent lamps. Where colour performance is important the sodium lamp, with its rather warm golden effect, may not be suitable and the choice is usually between the tubular fluorescent lamp and the metal halide lamp. A limitation of the fluorescent lamp is the restricted loading per point (i.e. more lamps are required per unit surface area) and in certain workshops where luminaire positioning at heights is required (in workshops with overhead travelling cranes for example) the high

Figure 7.1 Factory lighting where fine work is carried out and colour rendering is important, making use of reflector luminaires with tubular fluorescent lamps. (Courtesy Lighting Industry Federation)

pressure discharge lamp with its higher loading per point (generally up to 1 kW) is often selected. *Figure 7.1* shows factory lighting where fine work and colour rendering are important.

7.6 Illuminances

The illuminance (lighting level) required depends upon such things as the visual performance necessary for the tasks involved and general comfort and amenity requirements. The average illuminance out of doors in the UK is about 5000 lux on a cloudy day, but may be 10 times that on a sunny day. Inside a workplace, the illuminance from natural light at, say, a desk next to a window, will probably be only about 20% of the value obtaining outdoors. As working areas get further from windows the natural light produces illuminance values of perhaps only 1 to 10% of outdoor values so requires supplementing by artificial lighting. The normal way of expressing the effectiveness with which daylight reaches an interior is termed daylight factor[8].

In normal practice, decisions should be based on the recommendations of the Code for Interior Lighting produced by the Chartered Institution of Building Services Engineers[9] (CIBSE). Typical values of maintained illuminance for certain locations and tasks are given below but for detailed information, for particular industries and tasks, reference should be made to CIBSE[11]. Guidance and advice can be obtained from an HSE publication[10] and the Lighting Industry Federation[12] However, HSE requirements deal only with health and safety issues, whereas CIBSE recommendations also take account of cost effectiveness, productivity and amenity.

Although the term maintained illuminance represents levels that are good for general purposes, increases over the figures given may be necessary where tasks of high visual difficulty are undertaken, or low reflection or contrast are present, or where the location is a windowless interior. The CIBSE Code for Interior Lighting[9] gives criteria on which such adjustments can be based.

Location and task	*Standard service illuminance* (lx)
Storage areas, plant rooms, entrance halls etc.	150–200
Rough machinery and assembling, conference rooms, typing rooms, canteens, control rooms, wood machinery, cold strip mills, weaving and spinning etc.	300–400
Routine office work, medium machinery and assembly etc.	500
Spaces containing vdts used regularly as part of office tasks.	300–500
Demanding work such as in drawing offices, inspection of medium machinery etc.	750
Fine work requiring colour discrimination, textile processing, and fine machinery and assembly etc.	1000
Very fine work, e.g. hand engraving and inspection of fine machinery and assembly	1500

For a discussion on average maintained illuminances, minimum measured illuminances and for maximum ratio of illuminances between working and adjacent areas see reference 9.

7.6.1 Maintenance of lighting equipment

Dust, dirt and use will progressively reduce the light output of lamps and luminaires. Attention to good general cleaning and maintenance, and a realistic lamp replacement policy will help maintain the illuminance within recommendations. The expected maintenance regime is an essential factor in calculating the number of luminaires required for an installation. The maintenance regime appropriate to a building will depend on the activities carried out, the amount of dirt and dust carried in from outside and the type of lighting equipment in use. Some modern lamps lose light output much more slowly than older types, though luminaires will soil just as quickly.

7.7 Factors affecting the quality of lighting

The eye has the faculty of adjusting itself to various conditions and to discriminating between detail and objects. This visual capacity takes time to adjust to changing conditions as, for example, when leaving a brightly lit workroom for a darkened passage. Sudden changes of illuminance and excessive contrast between bright and dark areas of a workplace should be avoided.

A recent problem, resulting from the introduction of word-processors and other equipment using vdts, is the effect on eye discomfort and general well being of viewing screens for extended periods of time. Problems can be increased if the contrast between the screen and paper task is too great, if there is excessive contrast between the screen and background field of view, and if there are reflections of bright objects (luminaires, windows or even white shirts) in the screen. Lighting installations in such areas must comply with the requirements of the DSE Regulations[5]. The CIBSE has published specific guidance in this area[11] and there is also a luminaire certification scheme run by the Lighting Industry Federation[12].

7.7.1 Glare

Glare causes discomfort or impairment of vision and is usually divided into three aspects, i.e. disability glare, discomfort glare, and reflected glare.

It is referred to as disability glare if it impairs the ability to see clearly without necessarily causing personal discomfort. The glare caused by the undipped headlamps of an approaching car is an example of this.

Discomfort glare causes visual discomfort without necessarily impairing the ability to see and may occur from unscreened windows in bright sunlight or when over-bright or unshaded lamps in the workplace are significantly brighter than the surfaces against which they are viewed, e.g. the ceiling or walls.

Reflected glare, which can be disability glare or discomfort glare, is the effect of light reflected from a shiny or polished non-matt surface. The visual effect may be reduction of contrast, or distortion, and can be both irritating and, in certain workplaces, dangerous.

7.7.2 Glare indices

For many years in the UK, a glare index system has been in use for quantifying the effects of direct glare. It is also in use in certain other countries.

The relationship between direct glare discomfort and the factors affecting it, i.e. source brightness (B_s), background brightness (B_b), apparent size of the source angle (ω), and an index (p) representing the position of the source in relation to the direction of vision, is as follows:

$$\text{Glare index} = 10 \log_{10} \left[0.5 \times \text{constant} \sum \frac{B_s^{1.6} \, \omega^{0.8}}{B_b} \times \frac{1}{p^{1.6}} \right]$$

A set of comprehensive tables, based on this formula, has been produced by the CIBSE giving glare indices for a wide range of situations, types of luminaires etc., and these should be referred to for specific advice. Examples of numerical values of maximum desirable glare indices range from about 19 for a general office to about 28 for industrial work. Figures above the recommended levels for a given location may lead to visual discomfort.

Separate advice has been published[11] on reducing glare in premises where vdts are in use. This includes factories and workshops as well as offices. Forthcoming EU standards for lighting are likely to use the Unified Glare Rating system in place of the glare index. It is not yet known when these standards will come into force or how much influence they will have on UK practice.

7.7.3 Protection from glare

The most common cause of glare results from looking directly at unscreened lamps from normal viewing angles. Any form of diffuser or louvre fitted over the lamp, or a suitably placed reflector used as a screen will help to reduce the effect of glare from a lamp. The minimum screening angle below the horizontal should be about 20°, though greater angles are specified for areas containing vdts[11]. Reflected glare can only really be eliminated by changing the offending shiny surface for a matt one, or by adjusting the relative positions of light source, reflective surface and viewer.

Glare from sunlight coming through windows can be reduced by using exterior or interior blinds but this reduces the amount of natural lighting. It may be more effective to rearrange the workplace so that the windows are not in the normal direct field of view.

7.7.4 Effect of shadow

Shadow will affect the amount of illumination, and its impact on people in working areas will depend on the task being performed, and on the disposition of desks, work benches etc. The remedy is to use physically large luminaires (not necessarily with higher light outputs) or to increase their number. *Figure 7.2* illustrates factory lighting where the illuminance is to recommended standards.

7.7.5 Stroboscopic effect

The earlier type of tubular fluorescent lamp and discharge lamp were criticised because of the possibility of a stroboscopic effect. The light

Figure 7.2 Factory lighting of correct illuminance, free from shadow and glare, making use of high pressure discharge lamps. (Courtesy Thorn Lighting Ltd)

output from most lamps shows a cyclical variation with the alternating current, although in most circumstances this is not noticeable. However, it can cause a piece of rotating machinery to appear stationary or to be rotating slowly when, in fact, it is rotating at many times a second. This can be extremely dangerous. However, with modern fluorescent lamps and some discharge lamps the problem has been minimised by reducing the flicker effect. Where stroboscopic effects pose a particular danger they can be eliminated since it is possible to operate linear fluorescent and compact fluorescent lamps on electronic control gear at high frequency which both minimises the cyclic variation of light output and changes its frequency so that it is no longer visible as flicker. Alternatively, in most industrial and many commercial buildings it is possible to connect successive luminaires to the three phases of the power supply, which eliminates most flicker and stroboscopic effects.

7.7.6 Colour effect

The reflection of light falling on a coloured surface produces a coloured effect in which the amount of colour reflected depends upon the light source and the colour of the surface. For example, a red surface will only appear red if the incident light falling upon it contains red: under the almost monochromatic yellow of sodium street lighting, for example, a red surface will appear brown. The choice of lamp is important if colour effect or 'warm' or 'cool' effect is required and can be as important a consideration as the illuminance itself. Where accurate colour judgements have to be made the illuminance should be not less than 1000 lux and it may be appropriate to use either lamps whose colour rendering index is above 90 (CIE colour rendering group 1) or exceptionally special 'artificial day light' fluorescent lamps – commonly known as DE5 lamps.

7.8 Use of light measuring instruments

The human eye is unreliable as an indicator of how much light is present. For accurate results in the measurement of the illuminance at a surface it is necessary to use a reliable instrument. Light meters are available for this purpose.

A light meter, normally adequate for most locations, is a photocell which responds to light falling on it by generating a small electric current which deflects a pointer on a graduated scale measured in lux or, more commonly nowadays, causes a number to be displayed on a digital display. Most light meters have a correction factor built into their design to allow for using a filter when measuring different types of light (daylight, tubular fluorescent lamps, high pressure sodium lamps etc.). The recommended procedure for taking measurements with a light meter of this type is to:

1 Cover the cell with opaque material and alter the zero adjustment until the pointer reads zero on the scale.
2 Allow a few minutes for the instrument to 'settle down' before taking a reading. A longer period will be required if the light is provided by tubular fluorescent lamps or high pressure discharge lamps which have only just been switched on as they take time to reach full light output.
3 Select the appropriate scale on the instrument, i.e. that which gives the greatest deflection of the pointer or where the reading is closest to the upper end of the range.
4 If readings are to be taken during daylight two readings are necessary:
   (a) with the lights on and with the window blinds drawn back so as to record the combined effect of natural and artificial light, and
   (b) with the same natural light conditions as in (a) but with the artificial lights switched off.
   The result required, i.e. the measure of the artificial light, is the difference between the two readings. If the two readings are large and approximately equal it will be necessary to re-check the artificial light reading after dark.

The measured illuminance should be checked against the maintained illuminance for the location and task, taking account of the requirements, laid down by the CIBSE for the relevant areas[9]. The correct use of a light meter is an important aid to establishing good levels of lighting. However, to ensure accurate readings the instrument should be kept in its case when not in use and away from damp and excessive heat. It is also advisable to have the calibration checked by the manufacturer every year, though this is not cheap and it may be more cost effective to buy a new meter annually.

Do not overestimate the accuracy of the readings you obtain. Few hand-held meters are capable of measuring illuminance more accurately than within 10%, and the position of measurement can affect the measurement considerably. It is possible for measurements to differ from calculations by up to 60% for direct illumination and 20% for calculations involving interreflections. For maximum accuracy, measure at points on a regular grid through the space and average the results. Accuracy will be particularly suspect at low levels even if the meter itself has various ranges.

References

1. H. M. Government, *Workplace (Health, Safety and Welfare) Regulations 1992*, regulation 8, HMSO, London (1992)
2. H. M. Government, *Provision and Use of Work Equipment Regulations 1992*, regulation 21, HMSO, London (1992)
3. H. M. Government, *The Docks Regulations 1988*, regulation 6, HMSO, London (1988)
4. H. M. Government, *The Electricity at Work Regulations 1989*, regulation 15, HMSO, London (1989)
5. H. M. Government, *Health and Safety (Display Screen Equipment) Regulations 1992*, the schedule, HMSO, London (1992)
6. BS 6100, *Glossary of building and civil engineering terms, Section 3.4 Lighting*, BSI, London (1995), also International Commission on Illumination, publication 17.4, *International lighting vocabulary*, 4th edn, CIE-UK, c/o CIBSE, London (1987)
7. For details contact the Lighting Industry Federation, see below.
8. Building Research Establishment, Digest 309, *Estimating daylight in buildings, Part 1*; Digest 310, *Estimating daylight in buildings, Part 2*, CRC Ltd, London
9. The Chartered Institution of Building Services Engineers, *Code of Interior Lighting*, CIBSE, London (1984)
10. Health and Safety Executive, *Lighting at Work*, Health and Safety: Guidance Booklet No. HS(G)38, HSE Books, Sudbury (1989)
11. Chartered Institution of Building Services Engineers, *Lighting Guide 3. The visual environment for display screen equipment*, CIBSE, London (1996)
12. Lighting Industry Federation, Swan House, 207 Balham High Road, London SW 17 7BQ

Further reading

In addition to the above, numerous booklets and pamphlets on lighting for occupational premises and processes may be obtained from:
Chartered Institution of Building Services Engineers, 222 Balham High Road, London SW12 9BS. Relevant publications on specific types of premises include:
Lighting Guide 1, *The industrial environment* (1989)
Lighting Guide 2, *Hospitals and health care buildings* (1989)

Lighting Guide 4, *Sports* (1990)
Lighting Guide 5, *The visual environment in lecture, teaching and conference rooms* (1991)
Lighting Guide 7, *Lighting for offices* (1993)
Lighting Guide 8, *Lighting for museum and art galleries* (1994)
Technical memorandum 12: *Emergency lighting* (1986)
Guidance note 2. *Healthy workplaces: Guidance on complying with the 1992 health and safety regulations* (1993)
Building Research Establishment, Garston, Watford, Hertfordshire WD2 7JR. Publications available from: CRC Ltd, Bowling Green Lane, London EC1R 0DA
Lighting Industry Federation, Swan House, 207 Balham High Road, London SW17 7BQ

Chapter 8

Managing ergonomics

Nick Cook

8.1 Introduction

Ergonomics is about people. This is probably the single most important aspect of the subject. It's about different kinds of people; fat people, thin people, tall people, short people, bright people, not so bright people, young people, old people, male people and female people. And increasingly it will include disabled people. It is about taking all these different types of people and making sure their tools, their jobs and their work environments do not injure them. It's also about making sure they can do their work as comfortably and as efficiently as possible.

All these factors can make the management of ergonomics a daunting prospect. To do it cost effectively managers and health and safety professionals need a process for identifying and controlling ergonomic risk in the workplace. They need to know when to call in specialists and when to rely on their own in-house resources and commonsense.

This chapter aims to give a basic introduction to the subject which will help with the management process. Getting ergonomic management right is important: not only for employee health but also for the health of the business.

8.2 Ergonomics defined

Ergonomics is the reason why chairs are made with comfortable, adjustable backrests. It's the reason why VDU screens don't display pink letters on a magenta background and it's the reason why car controls are all in easy reach. And if it isn't, it should be.

A more formal definition was provided by Professor K. F. H. Murrell[1] in 1950 when he defined ergonomics as:

> 'The scientific study of the relationship between man and his environment.'

In truth there are probably almost as many definitions of ergonomics as there are practitioners. For example, in 1984 Clarke and Corlett[2] proposed the following definition:

> 'The study of human abilities and characteristics which affect the design of equipment systems and jobs . . . and its aims are to improve safety and . . . well being.'

Other definitions are very detailed indeed in their attempts to capture the essence of this wide ranging and evolving field. Christianson *et al.*[3] in 1988 defined ergonomics as:

> 'That branch of science and technology that includes what is known and theorised about human behavioural and biological characteristics that can be validly applied to the specification, design, evaluation, operation and maintenance of systems to enhance safe, effective and satisfying use by individuals, groups and organisations.'

Although no one could claim this definition is verbally ergonomic, it is comprehensive. It emphasises the range of human attributes (physical and mental) and the range of work attributes (the job and the equipment from design to maintenance) covered by ergonomics.

But perhaps the last word on the subject of definition should go to Britain's first Chief Medical Inspector of Factories, Sir Thomas Legge[4]. In the nineteenth century he proposed the following criteria for assessing work:

> 'Is the job fit for the worker and is the worker fit for the job?'

The field of ergonomics embraces a wide range of disciplines, from psychology to anatomy.

8.3 Ancient Egyptians and all that – a brief history of ergonomics

This section aims to put flesh on these definitions by giving some early practical examples of ergonomic issues and a brief outline of the development of the science.

The formal science of ergonomics may be relatively new but ergonomic issues have been around as long as humans. One of the earliest examples dates from over 10 000 years ago. Studies[5] on the female skeletons of Neolithic women who lived in what is now Syria showed specific deformities. These have been attributed to long hours spent kneeling down using a stone shaped rather like a rolling pin to crush corn on a saddle quern. This operation caused damage to the spine, neck, femur, arms and big toe (a result of the need to stabilise the kneeling position adopted for this job).

The recently excavated skeletons of Egyptian pyramid builders tell with grim eloquence of an ergonomic hell. Most of the skeletons show abnormal bony outgrowths (osteophytes) caused by manually dragging the 2.5 tonne blocks used to build the pyramids. Their bones also showed wear and tear while spines were damaged and some skeletons had severed limbs or splintered feet. Small wonder that the workers died between the ages of 30 and 35 whereas the nobility survived to between 50 and 60[6]. Neolithic chieftains and Egyptian Pharaohs had very little incentive to invent ergonomics since they could afford to adopt a 'pass me another worker this one is broken' approach.

It was with the industrial revolution that opportunities for ergonomic improvement really became apparent. Factories and mines in the nineteenth century were death traps. There were few safeguards on machines and workers, by and large relatively new to an industrial environment, were poorly trained to operate the machinery. In the new factories the emphasis was very much on work rate and long hours of work, both of which made workers susceptible to the hazards inherent in their labour.

In contrast with this, even though cottage industries were arguably exploited and overworked, handloom weavers or spinners at least had control over how and when they worked. Working in their own homes they could take breaks as and when they needed to.

Mining, a growth industry during the industrial revolution, was yet another ergonomic nightmare. Cornish tin miners would fall from exhaustion after climbing the ladders to the surface at the end of their shifts, tragically, often as they neared the top. Cages were made available as mines got too deep for manual ladders but even these were not safe. In the 1930s, George Orwell[7] wrote of the row of 'buttons' down miners' backs. These were the marks left by the always too low roof beams in the tunnels where miners, bent double as they moved along, would scrape their backs.

Ergonomics first began to be used with the work of F. W. Taylor[8] and F. B. Gilbreth[9] who conducted studies with the aim of increasing production efficiency rather than making the job less hazardous for employees. Their mission was to make work more scientific, to calculate the most efficient means of working using detailed timings of the physical movements made by individuals in the course of their work. Taylor's method focused on breaking down production work into simple functions and allocating each employee one specific task.

Taylor's philosophy became the basis of Ford's success with production lines but even at the time they were controversial enough to attract a congressional investigation. Taylor's attitude, and with it the attitude of this early approach to ergonomics, is perhaps best summed up by his reply[10] to a question concerning those workers unable to meet the demands of the stopwatch:

> 'Scientific management has no place for a bird that can sing and won't sing.'

It is perhaps not surprising that Henry Ford had to pay his workers twice the rate that other car companies (not yet using production line methods) paid. The studies did not allow for the effect on human beings of the sheer grinding monotony of production line work.

The science of ergonomics gained momentum during the Second World War. The complexity of aircraft, especially when fitted with equipment such as radar, led to confusion and fatigue among aircrew which in turn led to poor performance in an environment where the penalty for poor performance was likely to be very high.

In the early nineteenth century, a Polish scientist, Wojciech Jastrze-bowski, first coined a term similar to ergonomics (derived from the Greek *ergos* meaning laws and *nomos* meaning work), but the term did not really occur in common use until used by Professor K. F. H. Murrell, a founder member of the Ergonomics Society, in the middle of the twentieth century. In the USA the terms *human factors* or *human factors engineering* have been used, although the term ergonomics is being increasingly used. Today ergonomics still has important applications in the armed services and the aerospace industry but is being increasingly applied in the non-military working environment.

8.4 Ergonomics – has designs on you

Risk – any risk – is best controlled at source, thus protecting the working population as a whole. The risk to hearing from a noisy motor is best controlled by replacing the motor with a quieter one rather than supplying people with ear-muffs. And the risk of silicosis to workers doing grinding operations was reduced by making grinding wheels from carborundum rather than sandstone. In ergonomics the same principle is followed. Whether considering a job, the tools or the equipment needed to do it, the aim should always be to control the risk at the design stage. Ergonomics has many concepts and techniques to help achieve this goal. Some of the main ones are discussed below.

8.5 Ergonomic concepts

8.5.1 Usability

Usability is the capability of a system to be used safely and efficiently. The fact that all humans are different must be taken into account when assessing usability. For example, shorter stockier people are better able to deal with the G-forces experienced when executing tight turns in a fighter plane. This is the because their hearts have less work to do to get the blood to the head. If it is not possible to design planes or flying suits to eliminate the effects of G-forces then it may be necessary to select short stocky people to become fighter pilots, i.e. fit the person to the job. In general, however, it is more desirable to fit the job to the person as has occurred in the development of voice controlled word processing software for workers handicapped by repetitive strain injury (RSI).

In addition to the diversity of individuals likely to operate the system, the specific range of physical and environmental conditions must be specified. For example, are controls easily accessible? Is the room temperature and humidity satisfactory? The specific social and organisational structure should also be taken into account.

8.5.2 The human–machine interface

The human–machine interface is an imaginary boundary between the individual and the machine or equipment. When humans operate tools or machinery, information and energy have to cross this boundary. Consider a helicopter pilot. Information passes across the interface from the machine to the pilot via the control panel display. In response to this information energy then passes from the pilot to the machine via the controls. This example is the basic model for the interaction between humans and machines. It has been described as a closed loop system. The human receives the information from the machine, processes the information and responds by operating controls as appropriate. The machine responds to the controls and then sends information to the human via a display.

The ergonomic design of the interface (e.g. the controls and panel display) is very important. It has to fit the individual's physical and mental capabilities. Getting it wrong can be fatal. For example, the pilot of a British Airways helicopter that crashed into the sea off the Isles of Scilly claimed that he didn't see the warning light on the altimeter[11]. For people of his particular height the joystick obscured the view. Clearly human variability had not been taken into account for this particular human–machine interface. It was a costly oversight. Out of 26 people aboard the helicopter 20 died.

The following sections consider displays and controls, the two fundamental elements in the human–machine interface, in more detail.

8.5.2.1 Displays

The type of display must meet the needs of the human operating the machine or equipment and the display itself must be as clear and as easy to read as possible. It should not overload the operator with too much data but must take into account the information needed and how quickly it needs to be assimilated. The importance of getting this right is underlined by the fact that poor display design was a contributory factor to the nuclear power station incident at Three Mile Island.

The type of display should be appropriate to the data displayed. For example, analogue displays are better for showing rates of change. A needle on a dial or even a column of mercury in a thermometer gives a human operator a very clear picture of the rate of change of temperature. This will be much better than a digital display which will simply show a series of (possibly rapidly) changing numbers. Analogue displays are also very good at indicating whether the temperature remains within a desired range, especially if that range is marked on the gauge. A digital display on the other hand is very good where more precise readings are required. For example, provided the temperature is not varying too quickly, it is much easier to get an accurate reading from a digital display than from an analogue display.

A visual display may not be the most important way to present data. Using the example above, there may be serious consequences if the temperature strays from a pre-defined range. In this case an audible alarm might be needed to draw immediate attention to the divergent condition.

Much has been written about the relative merits of different types of displays and how they should be fitted to the needs of the operator. Grether and Baker[12] consider the preferred display by information type. Wilson and Rajan[13] in a comprehensive review of systems control consider the relative merits of panel and VDU displays. As far as the nuts and bolts of display criteria is concerned, one of the best introductions is the very practical account given by Cushman et al.[14]

An ergonomic term used in the context of information display is *coding* which is simply the way the information is represented. Letters or numbers can be used to represent elements within a system. The size of a symbol on a screen can be used to represent magnitude. Brightness could be used to represent temperature and colour could be used to help classify data. However, colour should only be used as redundant code, i.e. it should not be the only means of displaying the information, but some form of lettering or pictogram should be used to prevent the possibility of errors due to colour blindness. For example, before recent changes resulting from an EU directive, UK fire extinguishers were colour coded according to type. The type of extinguisher, however, was also indicated by text on the extinguisher.

8.5.2.2 Text clarity

A large amount of time is spent reading (or in some cases deciphering) text. It is surprising that writing style has been relatively neglected as an ergonomic issue. Text is a very important interface, not only between operators and machines and equipment (e.g. operating manuals) but also between humans and the organisation in which they work (e.g. work procedures, conditions of employment, policies etc.). At a more fundamental level, text is the interface between workers and the subject knowledge necessary for their jobs. But this relatively neglected area is beginning to find its rightful place in texts on ergonomics. A recent textbook on the methodology of ergonomics[15] has devoted a whole chapter to this important topic where one of the important concepts covered is *readability*.

A possible tool for evaluating readability is the Gunning Fog Index which works on the premise that long sentences and long words make text difficult to understand. In order to work out the Gunning Fog Index the following equation is used:

Gunning Fog Index = (average sentence length + number of long words) × 0.4

The average sentence length can be calculated by counting the number of sentences in 100 words of text and dividing 100 by that number. Long words in that sample of 100 words are any words that contain three or more syllables (excluding proper names).

The Gunning Fog Index gives a rough guide to the readability of the sample of text. A guide to the significance of specific Fog Indices is:

10   – would be readily understandable by the average 15-year-old secondary school pupil. (The *Daily Mail* is about 10.)
14–16 – would be readily understandable by university students.
>18   – the text is now becoming too difficult to understand without serious study.

Many academic texts and corporate documents have Fog Indices which greatly exceed 18 and it is almost as if the writers make their texts complicated in order to impress and confer authority on them. There is no excuse for this. Text should be as user friendly as possible and follow the KISS principle – Keep It Short and Simple.

There are other indices of readability (e.g. the Flesch Reading Ease Index). Many word processing packages will not only calculate a readability index on input text but also give guidance on other aspects of style.

An empirical rule of thumb is to try to keep the average sentence length to about 16 words although some sentences could be up to 26 words. Sentence lengths should be varied. These are not rules to be slavishly obeyed but will, if followed, generally help to improve the readability of most texts. While readability is an important factor, the text layout and type and range of font sizes should be chosen carefully to ensure that nothing in the appearance of the written text creates a mental barrier to the assimilation of the information.

8.5.2.3 Controls

Controls – e.g. levers, buttons, switches, foot pedals – represent the other half of the man–machine interface. To optimise a design it is necessary to take into account factors such as:

(a) *Speed:* Where a fast response is required the control should be designed so that it can be operated by the finger or the hand. The reason for this is that the hand and finger give the greatest speed and dexterity with the least effort.
(b) *Accuracy:* Using a mouse to select icons on a screen is an example of an operation where accuracy may be required. For such operations it is important to get the *control/display ratio* (C/D ratio) right. This is the ratio between the amount of movement of the control device to the degree of effect. In the case of a mouse it would be the ratio of the distance required to move the mouse to the resulting distance moved by the cursor on the screen. Where the C/D ratio is low a large movement of the mouse is required to achieve a relatively small movement of the cursor on the screen. This results in a slower but more accurate operation.

(c) *Force:* It should not be necessary to have to use excessive force in order to operate controls. On the other hand, a certain amount of force may be necessary to prevent a control being accidentally tripped.

(d) *Population stereotypes*[16]: This term refers to the expectations that controls work in certain ways. For example, we expect to have to turn the steering wheel of a car clockwise to go right and anticlockwise to go left. We expect the effect exercised by vertical levers to increase as we pull them towards the body and to decrease when pushed away. These stereotypes are the expectations of most of the population.

In some cases the stereotypes will differ from country to country. For example, to switch a light on in the UK the switch has to be moved to the down position. In the USA this convention is reversed. When a dial is used to alter a display the direction of movement on the display conventionally moves in the direction of the nearest point of the dial to the display. Dials are normally located below the display to leave the display in view when the dial is being operated. Therefore clockwise movement of a dial would be expected to move a pointer on a display to the right.

Conforming to the prevailing conventions dictated by stereotyping is desirable. There is always the danger that controls which do not conform may inadvertently be turned the wrong way. Population stereotypes also exist for displays. For example, red signifies danger while green signifies that it's safe to go.

8.5.3 Allocation of function

What proportion of a job should be done by the human operator and what proportion should be done by the machine? Hand tools are, inevitably, controlled by the user but for many machine tools automation is increasingly being employed.

To allocate function effectively, ergonomists need to consider the differences in capability between humans and machines. In an early attempt to allocate function at the design stage Fitts[17] listed the relative abilities of humans and machines. This list, subsequently modified by Singleton[18], included observations such as the fact that machines were much faster and more consistent than humans but that humans were a lot easier to reprogram and were by and large better at dealing with the unpredicted and the unpredictable. However, many have warned against the rigid application of such lists. The pace of technological change means that such lists will always be obsolete. Computer technology has made it possible to be more flexible on this issue. The allocation of function can be variable according to the type of person and operation[19].

8.5.4 Anthropometry

Anthropometry is the science of measurement of physical aspects of the human body. These include size, shape and body composition. Using data obtained from anthropometry, tools, equipment and workstations can be designed in such a way as to ensure the maximum comfort, safety and efficiency for individuals using them. Some anthropometric data is given in EN standards[20].

8.5.5 Error

How do we deal with human error in the workplace? Some employers take the attitude that as long as the equipment is working properly the blame for any errors must lie with the worker, possibly stemming from a lack of motivation, skill, training, or ability. This human error can be tackled by implementing a zero defects programme involving publicity posters, refresher training, job transfer and ultimately disciplinary action such as verbal warnings, written warnings, demotion and finally dismissal.

A more realistic approach is to assume that even the best people are prone to make mistakes. Jobs and equipment should be designed to work, not only from a functional point of view but also taking into account the fact that people can and do make mistakes. Equipment and methods of work should be designed to minimise the possibility and consequences of mistakes. Carried to extremes, the elimination of errors could lead to complete automation of the job dispensing with the operator altogether. Where this is not possible, reliance must be placed on interlocking and self-checking systems so that a wrong action or sequence of actions cannot be performed.

There is, thus, a need to assess human error at the design stage and one such approach is the technique of Human Reliability Assessment[21]. The aims of this technique can be summarised as:

(a) identify what can go wrong;
(b) quantify how often human error is likely to occur; and
(c) control the risk from human error either by preventing it in the first place or reducing its impact when it does occur.

Essentially, in health and safety terms this approach involves the assessment and control of risk where the steps to take include:

1 *Define the problem*: Basically this means looking at what work has to be done and identifying the ways in which mistakes by human beings may interfere or prevent this happening.
2 *Analyse the task*: What precise actions do humans have to carry out in order to do the job?
3 *Identify the human errors* and specify the ways in which they can be recovered. At each stage of the process under analysis, one technique[22] is to ask simple questions such as:
   (a) what if a required act is omitted?
   (b) what if required actions are carried out
       (i)   incorrectly?
       (ii) in the wrong order?
       (iii) too early?

(iv) too late?
(v) too much?
(vi) too little?
(c) what if a wrong act is carried out?

4 *Assess the probability of errors occurring and their likely seriousness.*
5 *Control*: Identify and implement steps to reduce the risk of human errors occurring. This could be done by improving human performance through training and/or improving the design of the work equipment and the environment.
6 *Review and audit*: Check that the control measures are effective. A quality assurance programme (such as ISO 9001) should be implemented to ensure the continuation of the error reduction programme.

For more detail on human reliability assessment see Kirwan[21].

8.6 Managing ergonomic issues in the workplace

This chapter has so far considered some of the steps that ergonomists might take during the design stage of processes and equipment, i.e. anticipating ergonomic problems and designing them out before they are inflicted on the workforce. Managers and health and safety professionals need to be aware of these issues when selecting equipment and machinery or managing the implementation of new processes.

It is equally important to have a procedure for managing ergonomics in an existing workplace. Very often managers and health and safety professionals inherit a workplace where, at the design stage, little if any thought had been given to ergonomic issues. Even where consideration has been given at the design stage improvements can still be made. Getting ergonomics right is an iterative process since often it is only after a process or piece of equipment has been used for some time that problems become apparent. All of which makes a good ergonomics programme that much more important.

The basic approach is one of risk assessment followed by action to eliminate any unacceptable risks and a maintenance and review process to ensure that the agreed work methods are followed. Workers should be informed of the reasons for, and given adequate training in, any necessary control measures. One of the keys to success is to involve the operators who carry out the work. Their involvement can result in risk assessments which are more realistic and controls which are practical and acceptable. Involvement also results in greater awareness of the health and safety issues by both managers and operators.

8.7 Work related upper limb disorders (WRULD)

8.7.1 Background

In October 1993 High Court Judge John Prosser dismissed a claim for industrial injury. Reuters journalist Rafiq Mughal[23] claimed to have contracted repetitive strain injury (RSI) while manning the Reuters Equities Desk. In dismissing the case, Judge Prosser said that 'RSI has no place in the medical textbooks'. He went on to refer to 'eggshell personalities who needed to get a grip on themselves'. Which was an unfortunate remark because if you have RSI it's extremely difficult to get a grip on anything.

However, there is still controversy over the exact meaning of the term RSI. In many cases patients suffer symptoms (in the case of Rafiq Mughal it was a tingling and numbness in his hands which eventually spread up his forearm and a painful right shoulder). But the precise causes are often not identifiable. Some doctors maintain that the term RSI should be reserved solely for this situation. Others, however, may use the term in the broader sense.

RSI as a term is a media rather than a medical invention. In the UK, *work related upper limb disorder* (WRULD) is used as a blanket term for a collection of disorders, most of which, unlike RSI, have symptoms which can be attributed to identifiable causes. In the USA the preferred term is *cumulative trauma disorder* (CTD).

8.7.2 Physiology of WRULD

WRULDs arise because of the physiology of the upper limbs. When performing an action with the hands it is the arm muscles that do the work. The force is transmitted to the bones in the fingers by a network of cables known as tendons. These tendons reach the fingers via the carpal tunnel, a bony arch located where the wrist joins the hand. In the course of moving a finger a tendon can travel as much as five centimetres. Tendons were well designed by evolution. To prevent wear and tear from friction each tendon is surrounded by synovial fluid which is contained in the synovial sheaf which covers the tendon in much the same way as electrical insulation covers a wire.

Superb as this system is it was never designed to cope with the stresses of modern work activities such as keyboard use. Mill[24] describes how such conditions can lead to injury, the main factor being *overuse*. Keyboard work, for example, requires an enormous amount of very quick, repetitive finger actions. These can lead to strain and injury particularly if carried out in awkward positions and for too long. Overuse can lead to quite specific injuries such as:

- *Tenosynovitis*: Inflammation of the synovial sheath.
- *Tendinitis*: Inflammation of a tendon.
- *Epicondylitis*: An inflammation of the tendons which attach muscle to the elbow. Depending on its exact nature, this condition is commonly called either tennis elbow or golfers elbow.
- *Carpal tunnel syndrome*: Tendons and nerves entering the hand have to squeeze through the carpal tunnel. This is an archway where the roof is made of cartilage and the sides and floor are composed of bones. Swelling of the synovial sheath surrounding the tendons puts pressure on the nerve and restricts the flow of blood into the hand causing

swelling of the wrist with pain, numbness and tingling which are typical symptoms.

Other specific conditions include *supraspinatus tendinitis* which affects the shoulder, *de Quervain's disease* which affects the tendon sheath in the thumb, and *ganglions*, small cysts which appear on the wrists. These specific conditions are often associated with modern keyboard use. Advice on assessing the risks of upper limb disorders is given in an HSE leaflet[25].

The effects of overuse can be made much worse by two contributory factors. The first of these is *static posture*. Problems for certain limbs may be caused by repetitive movements but for other parts of the body it is the absence of movement. While the fingers are pounding away on the keyboard other parts of the body, e.g. arms and shoulders are often held rigidly still. Muscles are continuously contracted, continuously under tension. This can restrict blood flow and causes nerves to become trapped and compressed.

The other contributory factor is *stress*. The human reaction to stress comes straight out of our hunter–gatherer past, i.e. the fight or flight syndrome. Hormones to stimulate physical action are poured into the blood and while this was very necessary for a stone age man facing a sabre tooth tiger, it is not so good for today's humans sitting rigidly in front of a word processor. This lack of physical activity to disperse the hormones round the body can have long-term medical effects.

8.7.3 Managing the risks from display screen equipment (DSE)

The first step in managing the risks from DSE is to develop a strategy and identify the actions that need to be taken. This can be achieved through writing a Company Display Screen Equipment Standard which should contain the following sections:

(a) *Management intent*: This should state the company policy on the use of workstations and that they must not pose a risk to the health of users. Also, that users will be given information, instruction and training so that they know the reason for control measures and how to implement them.
(b) *Performance expectations*: These concern what needs to be done and cover items such as risk assessment, training, provision of work breaks, eyesight testing, ensuring that new equipment and accessories comply with the relevant British Standards and considering the suitability of the software from the point of view of the health, safety and ability of the user. Guidance on this topic is available from the HSE[26,27].
(c) *Guidance*: This section should give any additional guidance deemed necessary to meet the performance expectations.
(d) *Appendices:* These should contain additional information, e.g. diagrams showing the correct posture to adopt at workstations. They should also contain examples of any forms used, e.g. the risk assessment checklist form.

A well-written standard provides staff with a ready source of reference and helps to ensure a consistent approach across the organisation. The very act of preparation helps to clarify what needs to be done. But note that the key words are *well written* (see the section on text clarity above). A badly written standard which confuses the reader can result in a reluctance on the part of employees to get involved in a DSE management programme.

Employee involvement is vital. It starts with the Purchasing Department ensuring that all equipment purchased conforms to the appropriate British Standards. Then continues with the installer ensuring that all the equipment operates satisfactorily. Finally the employees carrying out a risk assessment using a simple checklist to assess his or her workstation. The checklist should cover all the elements of the workstation, i.e. the chair, the desk, the keyboard, the pointing device (e.g. the mouse), the display screen itself and the general environment.

Any risks deemed to be high should be eliminated. This could be as simple as giving an employee a larger mouse, supplying a wrist rest for the keyboard, or the provision of a suitable footrest. The office politics of workstations should also be taken into account. A recent article by Steemson[28] argues for allocating workstations on the basis of the amount of space the work requires rather than on the basis of status. The biggest workstation should not necessarily go to the most senior person in the section.

The training of DSE users should emphasise the importance of following the recommended method for operating the equipment. The showing of a suitable video followed by a discussion with hands-on experience has proved effective. Suitable videos can be obtained from organisations such as RoSPA.

Finally the effectiveness of the measures taken should be reviewed and, if found wanting, should be modified to achieve the required standard. Routine workplace inspections should identify any workstations that do not comply. Similarly, the written standard and assessment pro-formas should be reviewed periodically, particularly whenever any changes of workplace equipment or software occur.

It should be stressed that WRULD is not specifically a keyboard disease. Anyone carrying out work requiring repetitive use of their upper limbs is potentially at risk. Other occupations at risk include chicken process workers, telephonists, mail sorters, signers for the deaf, fabric sewers and cutters and musicians. However, the procedure outlined above is applicable to all occupations where WRULD is a risk.

8.8 Back issues

8.8.1 Background

In 1994–1995 116 million working days were lost as certified sickness through back pain[29]. This is the biggest single cause of lost working time in the UK. The annual cost to industry is estimated at over £5 billion.

When managing the risks of damage to the lower back, which is the site of most back injuries, it is important to realise that the causes are not limited to heavy lifting or heavy repetitive work. They are often the result of something we do more and more of in our working lives – sitting down. Poor sitting posture can damage the discs, the joints between the vertebrae, and the ligaments which connect them. Sitting exerts about 40% more pressure on the discs in the lower back (the lumbar region) than standing. Leaning forward exacerbates this situation where the extra pressure on the discs increases from 40 to 90%. The pressure can rise even more when leaning forward to pick up something like a book or a heavy file.

It is much better to have seating arranged so that the angle between the back and the thighs is greater than 90°. This will help decrease the pressure on the discs and vertebrae in the lumbar region.

The insidious thing about bad posture is that its effects manifest themselves gradually. Because it takes less effort, slumping can often seem more comfortable than sitting properly, but all that extra pressure on the lower back has a cumulative effect. The pressure can cause the *annulus fibrosis* (the elastic tissue which makes up the outer ring on the disc) to become worn and tattered. It fails to contain the *nucleus pulposus* (the fluid 'plug' inside the disc) which then presses on the nerve. Even when the annulus fibrosis heals the fluid plug will remain out of position. This is what is termed a prolapsed disc. It is also called a slipped disc but, in fact, this is a misnomer since the invertabral discs cannot actually slip.

The second major cause of back pain arises from wear and tear on the joints between the vertebrae (the facet joints). These joints are lubricated by synovial fluid held in place by a cartilage sheath. Over the years wear and tear can roughen these joints, causing them to lose their flexibility. Proper exercise can help keep them lubricated thus prolonging their useful life.

Ligaments can become permanently stretched by poor posture which can result in extra pressure on joints and discs and a permanently distorted posture. The damage accumulates until a single action triggers a long lasting and painful condition. This action can be as simple as picking up a child, twisting to get a case off the back seat of a car or even pulling up a weed. Ironically it can be the single, simple action that gets the blame rather than the years of abuse which caused the cumulative damage.

Part of a good ergonomics management programme should be not only to look at the job but to encourage individuals to adopt the correct posture and to develop the muscles around the spine. An employee awareness programme could usefully include recommendations for a regular (and of course properly supervised) programme of exercise in a gymnasium.

Training is an important element in ergonomic management and increasingly equipment manufacturers are providing health and safety training packages specific to their own products. One such example is Vauxhall Motors Ltd which has produced a training package for the owners of its cars. It consists of a video and booklet[30] and emphasises the importance of adjusting the seat in order to provide adequate support for the lower back. For salesmen who spend much of their time driving company cars, getting posture right is very important. Review and refresher training may also be necessary because the correct posture does not, at first, feel as comfortable as sitting incorrectly.

More general guidance can be obtained from the HSE[31].

8.9 Managing the ergonomics of disability

Managing the ergonomic requirements of disabled workers poses a unique but rewarding challenge. The rewards include the resources saved by not having to dismiss disabled workers and recruit and retrain their replacements, and the retention of the experience and training already invested in the individual. An added incentive for the effective ergonomic management of disability comes from the obligations placed on employers by the Disability Discrimination Act 1995 (see section 8.10 below). This Act has been criticised as 'toothless' because it does not have an enforcing authority, relying instead on civil action brought by the claimant. Nevertheless it does provide a clear outline of employers' duties.

One potential barrier to the management of the ergonomics of disability is the general lack of awareness about the disability itself. In the absence of an in-house occupational health department this can pose a serious problem. Most disabilities have their own associated charities who are a ready source of often well-produced informative material. Additionally, many of these charities have experts who focus on the implications of employing people with specific disabilities. Another source of information is the Employers Forum on Disability (Tel. 0171 492 8460).

Disability is an issue which, increasingly, is attracting the attention of the health and safety press. An example of such coverage is the *Working Wounded* series of articles published in RoSPA's journal, *Occupational Safety & Health*. These gave examples of ergonomic adjustments which employers can make to help people with specific disabilities including:

- *Arthritis*[32]: Ramps for wheelchair users, stair lifts, toilets with wide doors. Special consideration should be given to emergency escape from buildings. Advice given with the Fire Precautions (Workplace) Regulations 1997 emphasises the need to consider the disabled when reviewing means of escape arrangements.
- *RSI*[33]: Voice recognition software makes possible hand-free operation of software packages such as Word for Windows. However, if keyboards are used, preventive steps such as the provision of ergonomic keyboards (e.g. the keyboard manufactured by PCD Maltron) could be considered.
- *Occupational deafness*[34]: Loop Systems which transmit speech straight to a hearing aid enabling speech to be heard above the extraneous noise. 'Minicoms' enable deaf people to type text messages to each other's screens using ordinary telephones.

- *Visual impairment*[35]: Speech synthesisers on computers which, for example, will say the words as the user moves the cursor round a Word for Windows document. Electronic Braille: pins rise and fall under a strip on the keyboard to create a Braille impression of the document on the screen. Magnification programs: to make screen text larger for those with visual impairment.

Financial help may be available to pay for special measures such as those outlined above. Details can be obtained from local DETR job centres.

8.10 Legal requirements

The main reasons for implementing an effective ergonomics management system should always be the health of the workforce and the wealth of the business and not merely legal compliance. Nevertheless an awareness of the legal requirements will provide useful guidance in formulating a management strategy. The relevant requirements are summarised below.

8.10.1 The Workplace (Health, Safety and Welfare) Regulations (1992)

- adequate lighting (reg. 8)
- adequate space (reg. 10)
- suitable workstations, including suitable seating (reg. 11).

8.10.2 The Provision and Use of Work Equipment Regulations 1992

- equipment to be suitable for the intended purpose (reg. 5)
- suitable guards and protective devices (reg. 11). Ergonomic principles are inherent in the advice given about reg. 11(2) and contained in HSE Guidance[36], namely the requirement for:
  (a) fixed enclosing guards
  (b) other guards or protecting devices
  (c) protection appliances (jigs, holders, push sticks etc.)
  (d) the provision of information, instruction training and supervision.
  The guidance makes it quite clear that in considering which combination of the above measures to adopt, they should be considered in strict hierarchical order from (a) to (c). In other words the best ergonomic solution (fixed guarding so that workers do not have to do anything themselves in order to be protected) must always be the first to be considered
- controls for work equipment must be clearly visible and identifiable (reg. 17)
- suitable and sufficient lighting (reg. 21).

8.10.3 The Personal Protective Equipment at Work Regulations 1992

- that PPE should only be worn as a last resort (reg. 4(2)). This requirement takes into account that wearing PPE is never as ergonomically desirable as not having to wear it at all because the hazard has been removed
- that PPE should take account of the ergonomic requirements and the state of health of people who wear it (reg. 4(3)(b)). The Guidance[37] accompanying these Regulations states that when selecting PPE, not only the nature of the job and the demands it places on the worker should be taken into account, but also the physical dimensions of the worker.

8.10.4 The Manual Handling Operations Regulations 1992

- employers, if possible, to avoid giving employees work which could result in injury (reg. 4(1)(a))
- employers to do a risk assessment on all work which involves carrying and which could represent a risk of injury to employees (reg. 4(1)(b))
- reduce the risk of injury from jobs where manual handling cannot be avoided and which could cause injury (reg. 4(1)(b)(ii)).

The Guidance[38] accompanying these Regulations contains an enormous amount of illustrative advice: it is virtually a practical ergonomics textbook in its own right.

8.10.5 The Health and Safety (Display Screen Equipment) Regulations 1992

- employers to assess and control the risks to users of display screen equipment workstations (reg. 2)
- the Guidance[26] and its schedule outline the physical considerations necessary to meet the Regulations
- the schedule also takes into account the mental stress which may result from using the software and in Schedule 4 stipulates that:
  - software must be easy to use and, where appropriate, adaptable to the level of knowledge or experience of the operators or user;
  - no quantitative or qualitative checking facility may be used without the knowledge of the operators or users;
  - the principles of software ergonomics should be applied in particular to human data processing;
  - systems should display information in a format and at a pace which are adapted to operators or users.

8.10.6 The Supply of Machinery (Safety) Regulations 1992

- From an ergonomic point of view the requirements of these Regulations eliminate at source a great many of the risks from machinery by requiring manufacturers to comply with the appropriate safety-of-machinery standard and confirm the fact in a declaration of conformity they issue with the equipment which they supply (see 25.1.3).

8.10.7 The Management of Health and Safety at Work Regulations 1992

- require that employers take account of the capabilities of employees when allocating work (reg. 11(1)).

8.10.8 The Reporting of Injuries, Diseases and Dangerous Occurrences Regulations 1995

- require an employer to report incidents of RSI resulting from work activity. This includes conditions such as cramp, tenosynovitis or carpal tunnel syndrome.

8.10.9 The Disability Discrimination Act 1995

- makes it unlawful for an employer to discriminate against a disabled employee (s. 4(1))
- requires an employer to make reasonable adjustments in order that a disabled person is not placed at a disadvantage compared to people who are not disabled (s. 6). These adjustments could be ergonomic as discussed above.

8.11 Conclusion

A chapter such as this cannot hope to provide examples of all factors which need to be considered in order to manage ergonomics. The subject is too vast. It encompasses psychology on the one hand, and occupational hygiene issues such as heat, noise and lighting on the other. Managing these issues effectively is as much a question of managing specialist consultants who may have to be called in to supplement 'in-house' expertise. Whether bought-in expertise is used will depend upon the complexity of the issue and the depth of knowledge required to deal with it.

Increasingly, software programs are being developed that facilitate the carrying out of risk assessments in-house. One such example is a software package[39] which uses a questionnaire and a force gauge (to measure the force needed for lifting, pulling or pushing operations). This package allows non-specialists to assess the risk from manual handling tasks and has been used successfully to train a workforce to carry out manual handling risk assessments on the work they do.

Effective ergonomics can be both operationally and cost effective making sense both from a health and safety and from a business point of view.

Acknowledgement

The author wishes to thank Mrs J. Steemson of the Royal Society for the Prevention of Accidents (RoSPA) for invaluable help and guidance with the research and writing of this chapter.

References

1. Murrell, K.F.H., *Ergonomics, Man and his Working Environment*, Chapman and Hall, London (1965)
2. Clark, T.S. and Corlett, E.N. *The Ergonomics of Workspaces and Machines: A Design Manual*, Taylor and Francis, London (1984)
3. Christiansen, J.M., Topmiller, D.A. and Gill, R.T., Human factors definitions revisited *Human Factors Society Bulletin*, **31**, 7–8 (1988)
4. Legge, Sir Thomas
5. Molleson, Theya, The eloquent bones of Abu Hureyra, *Scientific American*, August 1994, 60 (1994)
6. *New Scientist*, p.8, 20 January 1996
7. Orwell, George, *The Road to Wigan Pier*, Penguin Books, 24, ISBN 0–14–018238–1
8. Taylor, F.W., *What is Scientific Management? Classics in Management*, rev. edn, American Management Association, New York (1970)
9. Gilbreth, F.B., *Science in Management for the One Best Way to do Work, Classics in Management*, rev. edn, American Management Association, New York (1970)
10. Beynon, Huw, *Working for Ford*, 2nd edn, Penguin Books, 147, (1984) ISBN 0 14 022590 0
11. *The Guardian*, 24 February 1984
12. Grether, W.F. and Baker, C.A. Visual presentation of information, Ch. 3, van Cott and Kinkade (1972)
13. Wilson, R. and Rajan, J.A. Human–machine interfaces for systems control, Ch. 13, *Evaluation of Human Work*, eds Wilson, J.R. and Corlett, E.N., Taylor and Francis, 383 *et seq.* (1995)
14. Cushman, W.H. *et al.* Equipment Design, Ch. 3, *Ergonomic Design for People at Work*, eds Eggleton, E.M. and Rodgers, S.H., Eastman Kodak Company, New York, 91 (1983)
15. Hartley, J., Methods for evaluating text, Ch. 11, *Evaluation of Human Work*, eds Wilson, J.R. and Corlett, E.N., Taylor and Francis, 286 *et seq.* (1995)
16. Cushman, W.H. *et al.*, Equipment design Ch. 3, *Ergonomic Design for People at Work*, eds Eggleton, E.M. and Rodgers, S.H., Eastman Kodak Company, New York, 107 (1983)
17. Fitts, P.M., *Handbook of Experimental Psychology*, chapter on Engineering psychology and equipment design, John Wiley, London (1951)
18. Singleton, W.T., *Man–machine Systems*, Penguin Books, Hammondsworth (1974)
19. Clegg, C., Ravden, S., Corbett, M. and Johnson, G., Allocating functions in computer integrated manufacturing: a review and a new method, *Behaviour and Information Technology*, **8/3**, 175–190 (1989)
20. British Standards Institution, Harmonised European standards:
   BS EN 294, *Safety of machinery – Safety distances to prevent danger zones being reached by the upper limbs* (1992)
   BS EN 349, *Safety of machinery – Minimum gaps to avoid crushing of parts of the human body* (1993)
   BS EN 614–1, *Safety of machinery – Ergonomic design principles – Part 1: Terminology and general principles* (1995)
   BS EN pr 999, *Safety of machinery – The positioning of protective equipment in respect of approach speeds of parts of the human body* (to be published)

21. Kirwan, B., Human reliability assessment, Ch. 31, *Evaluation of Human Work*, eds Wilson, J.R. and Corlett, E.N., Taylor and Francis, 921 (1995)
22. Swain and Guttman, *A Handbook of Human Reliability Analysis with Emphasis on Nuclear Power Plant Applications*. Nureg/CR-1278, USNRC, Washington, DC (1983)
23. Rafdiq Mughal *v.* Reuters
24. Mill, W.C., *RSI*, Thorsons, London (1994) ISBN 0 7225 2919 8
25. Health and Safety Executive, Leaflet Pack No. INDG 171, *Work related upper limb disorders: Assessing the risk*, HSE Books, Sudbury (1995)
26. Health and Safety Executive, Legal Series booklet No. L26, *Display Screen Equipment at Work. The Health and Safety (Display Screen Equipment) Regulations 1992. Guidance on the Regulations*, HSE Books, Sudbury (1992)
27. Health and Safety Executive, Leaflet Pack No. IND 36, *Working with VDUs*, HSE Books, Sudbury (1998)
28. Steemson, J., Space Craft, *Occupational Safety and Health*, October (1997)
29. Department of Social Security, Statistical Unit. Statistics of certified incapacity
30. Vauxhall Motors Limited, *Are you Sitting Comfortably?*, video and booklet, Vauxhall Motors Limited, Luton (1997)
31. Health and Safety Executive, Leaflet Pack No. INDG 242, *In the driving seat*, HSE Books, Sudbury (1997)
32. Cook, N. The working wounded: arthritis and rheumatism, *Occupational Safety and Health*, April, 17 (1995)
33. Cook, N., The working wounded: RSI, *Occupational Safety and Health*, August 36 (1996)
34. Cook, N., The working wounded: hearing impairment. *Occupational Safety and Health*, January, 28 (1996)
35. Cook, N., The working wounded: visual impairment, *Occupational Safety and Health*, June, 43 (1995)
36. Health and Safety Executive, Legal Series booklet No. L22, *Work equipment. Provision and Use of Work Equipment Regulations 1992 – Guidance on the Regulations*, HSE Books, Sudbury (1992)
37. Health and Safety Executive, Legal Series booklet No. L25, *Personal protective equipment at work. Personal Protective Equipment at Work Regulations 1992 – Guidance on the Regulations*, HSE Books, Sudbury (1992)
38. Health and Safety Executive, Legal Series booklet No. L23, *Manual handling. Manual Handling Operations Regulations 1992 – Guidance on the Regulations*, HSE Books, Sudbury (1992)
39. Human Focus Ltd, Human Focus and Ergonomic Design Consultants, tel: 0181 640 3535

General reading

Dul, J. and Weerdmeester, B., *Ergonomics for Beginners: A Quick Reference Guide*, Taylor and Francis. Full of clear, practical advice on doing risk assessments.
Chalmers Mill, W., *Repetitive Strain Injury*, Thorsons Health. A very clear introduction to RSI and its prevention.
Health and Safety Executive, Guidance booklet No. HSG 121, *A pain in your workplace? Ergonomic problems and solutions*, HSE Books, Sudbury (1994)
Wilson, J.R. and Corlett, E.N. (eds), *Evaluation of Human Work*, 2nd Edition, Taylor and Francis (1995). An extremely comprehensive survey of the methodology of ergonomics. Very academic and possibly of more value to specialists than industrial health and safety professionals.
Rogers, S.H. and Eggleton, E.M. (eds), *Ergonomic Design for People at Work* (2 volumes), The Human Factors Section, Eastman Kodak, published by van Rostrand Reinhold Company, New York (1983) Despite its age this is an excellent, clearly written and comprehensive survey of the subject. Written with practical health and safety professionals in mind.
Office Health and Safety: a guide to risk prevention, Unison, 1 Mabledon Place, London WC1H 9AJ. A clear and well-written brief guide for union members and safety representatives

Useful contacts

The Ergonomics Society, Devonshire House, Devonshire Square, Loughborough, Leicester-
    shire LE11 3DW. Tel: 01509 234904. Publishes a register of consultancy firms in their
    membership.
Ergonomics Information Analysis Centre, School of Manufacturing and Mechanical
    Engineering, University of Birmingham, Birmingham B15 2TT, Tel: 0121 414 4239
The Computability Centre: a national charity which aims to help workers disabled by RSI
    return to work. PO Box 94, Warwick CV34 5WS.

The Institution of Occupational Safety and Health

The Institution of Occupational Safety and Health (IOSH) is the leading professional body in the United Kingdom concerned with matters of workplace safety and health. Its growth in recent years reflects the increasing importance attached by employers to safety and health for all at work and for those affected by work activities. The Institution provides a focal point for practitioners in the setting of professional standards, their career development and for the exchange of technical experiences, opinions and views.

Increasingly employers are demanding a high level of professional competence in their safety and health advisers, calling for them to hold recognised qualifications and have a wide range of technical expertise. These are evidenced by Corporate Membership of the Institution for which proof of a satisfactory level of academic knowledge of the subject reinforced by a number of years of practical experience in the field is required.

Recognised academic qualifications are an accredited degree in occupational safety and health or the Diploma Part 2 in Occupational Safety and Health issued by the National Examination Board in Occupational Safety and Health (NEBOSH). For those assisting highly qualified OSH professionals, or dealing with routine matters in low risk sectors, a Technician Safety Practitioner (SP) qualification may be appropriate. For this, the NEBOSH Diploma Part 1 would be an appropriate qualification.

Further details of membership may be obtained from the Institution.

Appendix 2

Reading for Part I of the NEBOSH Diploma examination

The following is suggested as reading matter relevant to Part 1 of the NEBOSH Diploma examination. It should be complemented by other study.

Module 1A: The management of   Chapters   2.1–all
                     risk   2.2–paras. 8–11
       2.3–all
       2.4–paras. 1–3
       3.8–paras. 1–6
       4.7–para. 11

Module 1B: Legal and organisational   Chapters   1.1–all
       factors   1.2–all
       1.3–paras. 1–6
       1.7–para. 2
       1.8–all
       2.2–paras. 13 and 14
       2.6–paras. 1–4

Module 1C: The workplace   Chapters   1.7–para. 2
       3.6–all
       3.7–all
       4.2–all
       4.4–paras. 1–8
       4.6–paras. 2 and 4
       4.7–paras. 1, 2, 7 and 11

Module 1D: Work equipment   Chapters   4.3–all
       4.4–all
       4.5–all

Module 1E: Agents   Chapters   **3.1–all**
       3.2–all
       3.3–all
       3.5–paras. 1–6
       3.6–all
       3.8–paras. 4–7
       4.7–paras. 1–4

Module 1CS: Common skills   Chapter   2.5–para. 7

Additional information in summary form is available in *Health and Safety ... in brief* by John Ridley published by Butterworth-Heinemann, Oxford (1998).

List of abbreviations

ABI	Association of British Insurers
AC	Appeal Court
ac	Alternating current
ACAS	Advisory, Conciliation and Arbitration Service
ACGIH	American Conference of Governmental Industrial Hygienists
ACoP	Approved Code of Practice
ACTS	Advisory Committee on Toxic Substances
ADS	Approved dosimetry service
AFFF	Aqueous film forming foam
AIDS	Acquired immune deficiency syndrome
ALA	Amino laevulinic acid
All ER	All England Law Reports
APAU	Accident Prevention Advisory Unit
APC	Air pollution control
BATNEEC	Best available technique not entailing excessive costs
BLEVE	Boiling liquid expanding vapour explosion
BOD	Biological oxygen demand
BPEO	Best practicable environmental option
Bq	Becquerel
BS	British standard
BSE	Bovine spongiform encephalopathy
BSI	British Standards Institution
CBI	Confederation of British Industries
cd	Candela
CD	Consultative document
CDG	The Carriage of Dangerous Goods by Road Regulations 1996
CDG-CPL	The Carriage of Dangerous Goods by Road (Classification, Packaging and Labelling) and Use of Transportable Pressure Receptacle Regulations 1996
CDM	The Construction (Design and Management) Regulations 1994
CEC	Commission of the European Communities

CEN	European Committee for Standardization of mechanical items
CENELEC	European Committee for Standardisation of electrical items
CET	Corrected effective temperature
CFC	Chlorofluorocarbons
CHASE	Complete Health and Safety Evaluation
CHAZOP	Computerised hazard and operability study
CHIP 2	The Chemical (Hazard Information and Packaging for Supply) Regulations 1994
Ci	Curie
CIA	Chemical Industries Association
CIMAH	The Control of Industrial Major Accident Hazards Regulations 1984
CJD	Creutzfeldt–Jacob disease
COD	Chemical oxygen demand
COMAH	The Control of Major Accident Hazards Regulations (proposed)
COREPER	Committee of Permanent Representatives (to the EU)
COSHH	The Control of Substances Hazardous to Health Regulations 1994
CPA	Consumer Protection Act 1987
CTD	Cumulative trauma disorder
CTE	Centre tapped to earth (of 110 V electrical supply)
CWC	Chemical Weapons Convention

dB	Decibel
dBA	'A' weighted decibel
dc	Direct current
DETR	Department of the Environment, Transport and the Regions
DG	Directorate General
DNA	Deoxyribonucleic acid
DO	Dangerous occurrence
DSE(R)	The Health and Safety (Display Screen Equipment) Regulations 1992
DSS	Department of Social Services
DTI	Department of Trade and Industry

EA	Environmental Agency
EAT	Employment Appeals Tribunal
ECJ	European Courts of Justice
EC	European Community
EEA	European Economic Association
EEC	European Economic Community
EcoSoC	Economic and Social Committee
EHRR	European Human Rights Report

EINECS	European inventory of existing commercial chemical substances
ELF	Extremely low frequency
ELINCS	European list of notified chemical substances
EMAS	Employment Medical Advisory Service
EN	European normalised standard
EP	European Parliament
EPA	Environmental Protection Act 1990
ERA	Employment Rights Act 1996
ESR	Essential safety requirement
EU	European Union
eV	Electronvolt
EWA	The Electricity at Work Regulations 1989

FA	Factories Act 1961
FAFR	Fatal accident frequency rate
FMEA	Failure modes and effects analysis
FPA	Fire Precautions Act 1971
FSLCM	Functional safety life cycle management
FTA	Fault tree analysis

| GEMS | Generic error modelling system |
| Gy | Gray |

HAVS	Hand-arm vibration syndrome
HAZAN	Hazard analysis study
HAZCHEM	Hazardous chemical warning signs
HAZOP	Hazard and operability study
hfl	Highly flammable liquid
HIV+ve	Human immune deficiency virus positive
HL	House of Lords
HMIP	Her Majesty's Inspectorate of Pollution
HSC	The Health and Safety Commission
HSE	The Health and Safety Executive
HSI	Heat stress index
HSW	The Health and Safety at Work, etc. Act 1974
Hz	Hertz

IAC	Industry Advisory Committee
IBC	Intermediate bulk container
ICRP	International Commission on Radiological Protection
IEC	International Electrotechnical Committee (International electrical standards)
IEE	Institution of Electrical Engineers
IOSH	Institution of Occupational Safety and Health
IPC	Integrated polluton control
IQ	Intelligence quotient
IRLR	Industrial relations law report
ISO	International Standards Organisation
ISRS	International Safety Rating System

JHA	Job hazard analysis
JP	Justice of the Peace
JSA	Job Safety Analysis
KB	King's Bench
KISS	Keep it short and simple
LA	Local Authority
LEL	Lower explosive limit
$L_{EP.d}$	Daily personal noise exposure
LEV	Local exhaust ventilation
LJ	Lord Justice
LOLER	Lifting Operations and Lifting Equipment Regulations 1998
LPG	Liquefied petroleum gas
LR	Lifts Regulations 1997
lv/hv	Low volume high velocity (extract system)
mcb	Miniature circuit breaker
MEL	Maximum exposure limit
MHOR	The Manual Handling Operations Regulations 1992
MHSW	The Management of Health and Safety at Work Regulations 1992
MOSAR	Method organised for systematic analysis of risk
MPL	Maximum potential loss
M.R.	Master of the Rolls
NC	Noise criteria (curves)
NDT	Non-destructive testing
NEBOSH	National Examination Board in Occupational Safety and Health
NI	Northern Ireland Law Report
NIHH	The Notification of Installations Handling Hazardous Substances Regulations 1982
NIJB	Northern Ireland Judgements Bulletin (Bluebook)
NLJ	Northern Ireland Legal Journal
NONS	The Notification of New Substances Regulations 1993
npf	Nominal protection factor
NR	Noise rating (curves)
NRA	National Rivers Authority
NRPB	National Radiological Protection Board
NZLR	New Zealand Law Report
OJ	Official journal of the European Community
OECD	Organisation for Economic Development and Co-operation
OES	Occupational exposure standard
OFT	Office of Fair Trading
OR	Operational research

P4SR	Predicted 4 hour sweat rate
Pa	Pascal
PAT	Portable appliance tester
PC	Personal computer
PCB	Polychlorinated biphenyl
PHA	Preliminary hazard analysis
PMNL	Polymorphonuclear leukocyte
PPE	Personal protective equipment
ppm	Parts per million
ptfe	Polytetrafluoroethylene
PTW	Permit to work
PUWER	The Provision and Use of Work Equipment Regulations 1998
PVC	Polyvinyl chloride
QA	Quality assurance
QB	Queen's Bench
QMV	Qualifies majority voting
QUENSH	Quality, environment, safety and health management systems
r.	A clause or regulations of a Regulation
RAD	Reactive airways dysfunction
RCD	Residual cirrent device
RGN	Registered general nurse
RIDDOR	The Reporting of Injuries, Diseases and Dangerous Occurrences Regulations 1995
RM	Resident magistrate
RoSPA	Royal Society for the Prevention of Accidents
RPA	Radiation protection adviser
RPE	Respiratory protective equipment
RPS	Radiation protection supervisor
RR	Risk rating
RRP	Recommended retail price
RSI	Repetitive strain injury
s.	Clause or section of an Act
SAFed	Safety Assessment Federation
SC	Sessions case (in Scotland)
Sen	Sensitizer
SEN	State enrolled nurse
SIESO	Society of Industrial Emergency Services Officers
Sk	Skin (absorption of hazardous substances)
SLT	Scottish Law Times
SMSR	The Supply of Machinery (Safety) Regulations 1992
SPL	Sound pressure level
SRI	Sound reduction index
SRN	State registered nurse
SRSC	The Safety Representatives and Safety Committee Regulations 1977

SSP	Statutory sick pay
Sv	Sievert
SWL	Safe working load
SWORD	Surveillance of work related respiratory diseases
TLV	Threshold Limit Value
TUC	Trades Union Congress
TWA	Time Weighted Average
UEL	Upper explosive limit
UK	United Kingdom
UKAEA	United Kingdom Atomic Energy Authority
UKAS	United Kingdom Accreditation Service
v.	versus
VAT	Value added tax
VCM	Vinyl chloride monomer
vdt	Visual display terminal
VWF	Vibration white finger
WATCH	Working Group on the Assessment of Toxic Chemicals
WBGT	Wet bulb globe temperature
WDA	Waste Disposal Authority
WHSWR	The Workplace (Health, Safety and Welfare) Regulations 1992
WLL	Working load limit
WLR	Weekly Law Report
WRULD	Work related upper limb disorder
ZPP	Zinc protoporphyrin

Organisations providing safety information

Institution of Occupational Safety and Health, The Grange, Highfield Drive, Wigston, Leicester LE18 1NN 0116 257 3100

National Examination Board in Occupation Safety and Health, NEBOSH, 5 Dominus Way, Meridian Business Park, Leicester LE3 2QW 0116 263 4700 Fax 0116 282 4000

Royal Society for the Prevention of Accidents, Edgbaston Park, 353 Bristol Road, Birmingham B5 7ST 0121 248 2222

British Standards Institution, 389 Chiswick High Road, London W4 4AL 0181 996 9000

Health and Safety Commission, Rose Court, 2 Southwark Bridge, London SE1 9HS 0171 717 6600

Health and Safety Executive, Enquiry Point, Magnum House, Stanley Precinct, Trinity Road, Bootle, Liverpool L20 3QY 0151 951 4000 or any local offices of the HSE

HSE Books, PO Box 1999, Sudbury, Suffolk CO10 6FS 01787 881165

Employment Medical Advisory Service, Daniel House, Trinity Road, Bootle, Liverpool L20 3TW 0151 951 4000

Institution of Fire Engineers, 148 New Walk, Leicester LE1 7QB 0116 255 3654

Medical Commission on Accident Prevention, 35–43 Lincolns Inn Fields, London WC2A 3PN 0171 242 3176

The Asbestos Information Centre Ltd, PO Box 69, Widnes, Cheshire WA8 9GW 0151 420 5866

Chemical Industry Association, King's Building, Smith Square, London SW1P 3JJ 0171 834 3399

Institute of Materials Handling, Cranfield Institute of Technology, Cranfield, Bedford MK43 0AL 01234 750662

National Institute for Occupational Safety and Health, 5600 Fishers Lane, Rockville, Maryland, 20852, USA

Noise Abatement Society, PO Box 518, Eynsford, Dartford, Kent DA4 0LL 01322 862789

Home Office, 50 Queen Anne's Gate, London SW1A 9AT 0171 273 4000
Fire Services Inspectorate, Horseferry House, Dean Ryle Street, London
SW1P 2AW 0171 217 8728

Department of Trade and Industry: all Departments on 0171 215 5000
Consumer Safety Unit, General Product Safety: 1 Victoria Street,
London SW1H 0ET
Gas and Electrical Appliances: 151 Buckingham Palace Road, London
SW1W 9SS
Manufacturing Technology Division, 151 Buckingham Palace Road,
London SW1W 9SS

Department of Transport
Road and Vehicle Safety Directorate, Great Minster House, 76 Marsham
Street, London SW1P 4DR 0171 271 5000

Advisory, Conciliation and Arbitration Service (ACAS), Brandon House,
180 Borough High Street, London SE1 1LW 0171 396 5100

Health Education Authority, Hamilton House, Mabledon Place, London
WC1 0171 383 3833

National Radiological Protection Board (NRPB), Harwell, Didcot, Oxford-
shire OX11 0RQ 01235 831600

Northern Ireland Office
Health and Safety Inspectorate, 83 Ladas Drive, Belfast BT6 9FJ 01232
701444
Agricultural Inspectorate, Dundonald House, Upper Newtownwards
Road, Belfast BT4 3SU 01232 65011 ext: 604
Employment Medical Advisory Service, Royston House, 34 Upper
Queen Street, Belfast BT1 6FX 01232 233045 ext: 58

Commission of the European Communities, Information Office, 8
Storey's Gate, London SW1P 3AT 0171 222 8122

British Safety Council, National Safety Centre, Chancellor's Road,
London W6 9RS 0171 741 1231/2371

Confederation of British Industry, Centre Point, 103 New Oxford Street,
London WC1A 1DU 0171 379 7400

Safety Assessment Federation (SAFed), Nutmeg House, 60 Gainsford
Street, Butler's Wharf, London SE1 2NY 0171 403 0987

Railway Inspectorate, Rose Court, 2 Southwark Bridge, London SE1 9HS
0171 717 6630

Inspectorate of Pollution, Romney House, 43 Marsham Street, London
SW1P 3PY 0171 276 8083

Back Pain and Spinal Injuries Association, Brockley Hill, Stanmore,
Middlesex 0181 954 0701

Appendix 5

List of Statutes, Regulations and Orders

Note: This list covers all four volumes of the Series. Entries and page numbers in bold are entries specific to this volume. The prefix number indicates the volume and the suffix number the page in that volume.

List of Cases

Note: This list covers all four volumes of the Series. Entries and page numbers in bold are entries specific to this volume. The prefix number indicates the volume and the suffix number the page in that volume.

Ollett v. Bristol Aerojet Ltd (1979) 3 All ER 544, *1.144*

Page v. Freight Hire Tank Haulage Ltd (1980) ICR 29; (1981) IRLR
    13, *1.94*
Paris v. Stepney Borough Council (1951) AC 367, *1.149*
Parsons v. B.N.M. Laboratories Ltd (1963) 2 All ER 658, *1.80*
Pickstone v. Freeman plc (1989) 1 AC 66, *1.30*
Pitts v. Hill and Another (1990) 3 All ER 344, *1.146*
Planche v. Colburn (1831) 8 Bing 14, *1.80*
Polkey v. A.E. Dayton (Services) Ltd (1988) IRLR 503; (1987) All ER
    974, HE (E), *1.101, 1.108*

Queensway Discount Warehouses v. Burke (1985) BTLC 43, *1.115*
Quintas v. National Smelting Co. Ltd (1961) 1 All ER 630, *1.138*

R. v. Bevelectric (1992) 157 JP 323, *1.117*
R. v. British Steel plc (1995) ICR 587, *1.37*
R. v. Bull, *The Times*, 4 December 1993, *1.118*
R. v. George Maxwell Ltd (1980) 2 All ER 99, *1.8*
R. v. Kent County Council (6 May 1993, unreported), *1.118*
R. v. Secretary of State for Transport v. Factortame Ltd C 221/89; (1991)
    1 AC 603; (1992) QB 680, *1.26, 1.27*
R. v. Sunair Holidays Ltd (1973) 2 All ER 1233, *1.117*
R. v. Swan Hunter Shipbuilders Ltd and Telemeter Installations Ltd
    (1981) IRLR 403, *4.163–8*
Rafdiq Mughal v. Reuters (1993), *3.185–4*
Readmans Ltd v. Leeds City Council (1992) COD 419, *1.25*
Ready-mixed Concrete (South East) Ltd v. Minister of Pensions and
    National Insurance (1968) 1 All ER 433, *1.81*
Roberts v. Leonard (1995) 159 JP 711, *1.117*
Rowland v. Divall (1923) 2 KB 500, *1.84*
R.S. Components Ltd v. Irwin (1973) IRLR 239, *1.109*
Rylands v. Fletcher (1861) 73 All ER Reprints N. 1, *1.142*

Sanders v. Scottish National Camps Association (1980) IRLR 174, *1.109*
Scammell v. Ouston (1941) All ER 14, *1.78*
SCM (UK) Ltd v. W. J. Whittle and Son Ltd (1970) 2 All ER 417, *1.145*
Scott v. London Dock Company (1865) 3 H and C 596, *1.139*
Shepherd v. Firth Brown (1985) unreported, *1.141*
Sillifant v. Powell Duffryn Timber Ltd (1983) IRLR 91, *1.101*
Smith v. Baker (1891) AC 325, *1.158*
Smith v. Crossley Bros. Ltd (1951) 95 Sol. Jo. 655, *1.153*
Smith v. Leach Brain & Co. Ltd (1962) 2 WLR 148, *1.150*
Smith v. Stages (1989) 1 All ER 833, *1.137*
Spartan Steel and Alloys Ltd v. Martin and Co. (Contractors) Ltd (1972)
    3 All ER 557, *1.145*
Spencer v. Paragon Wallpapers Ltd (1976) IRLR 373, *1.103*
Stevenson, Jordan and Harrison v. Macdonald & Evans (1951) 68 R.P.C.
    190, *1.81*
Systems Floors (UK) Ltd v. Daniel (1982) ICR 54; (1981) IRLR 475, *1.82*

Series Index

Note: This index covers all four volumes of the Series. Entries and page numbers in bold are entries specific to this volume. The prefix number indicates the volume and the suffix number the page in that volume.